P9-DFP-613

"Suddenly science is close to understanding the Indian rope trick by which life emerged from non-life four billion years ago. Adam Rutherford has written an engaging account of both the mystery and its impending resolution; he has also provided a fascinating glimpse of the impending birth of a new, synthetic biology."

—Matt Ridley, author of *Genome*

"Rutherford makes his case with contagious enthusiasm. . . . Genuinely amazing biology is in the works, and Rutherford delivers a fascinating overview." —*Kirkus Reviews* (starred review)

"Engaging . . . May it augur many more top-drawer science books by Rutherford." —*Booklist* (starred review)

"Prepare to be astounded. There are moments when this book is so gripping it reads like a thriller." —*The Mail on Sunday* (UK)

"Fascinating . . . The extraordinary science and his argument are worth every reader's scrutiny." —*The Sunday Telegraph* (UK)

"Rutherford's academic background in genetics gives him a firm grasp of the intricacies of biochemistry—and he translates these superbly into clear English." —*Financial Times*

About the Author

Adam Rutherford is a science writer and broadcaster. Formerly an editor at *Nature*, he now presents BBC Radio 4's flagship weekly program, *Inside Science*, as well as a host of other documentaries. A geneticist by training, he has a PhD from University College London.

CREATION

How Science Is
Reinventing Life Itself

Adam Rutherford

CURRENT

CURRENT
Published by the Penguin Group
Penguin Group (USA) LLC
375 Hudson Street
New York, New York 10014

USA | Canada | UK | Ireland | Australia | New Zealand | India | South Africa | China
penguin.com
A Penguin Random House Company

First published in the United States of America by Current, a member of Penguin Group (USA) Inc., 2013
This paperback edition published 2014

First published in Great Britain in a different format and as
Creation: The Origin of Life / The Future of Life by Penguin Books Ltd.

THE LIBRARY OF CONGRESS HAS CATALOGED THE HARDCOVER EDITION AS FOLLOWS:
Rutherford, Adam, Ph. D.
Creation : how science is reinventing life itself / Adam Rutherford.
p. ; cm.
Includes bibliographical references and index.
ISBN 978-1-61723-005-9 (hc.)
ISBN 978-1-61723-011-0 (pbk.)
I. Title.
[DNLM: 1. Biogenesis—Popular Works. QH 325]
QH325.R818 2013
576.8'3—dc23   2013013441

Printed in the United States of America
10 9 8 7 6 5 4 3 2 1

Designed by Nancy Resnick

For David Rutherford, from whose cells I came

Contents

CREATION

Introduction

magine you were unlucky enough to get a paper cut opening this book. It's an annoying but trivial injury, painful but easily mended. Yet the response that this incision triggers is complex, organized, and profound. It's comparable to the human reaction to a large-scale catastrophe such as a flood or an earthquake. As in those disasters, the first phase is an emergency response.

Everything that occurs in and around your cut happens as a beautiful orchestration of individual living cells. At the precise moment the sharp edge of the paper slices through the outermost surface of your skin, cells embedded throughout your flesh, called nociceptors, spark into action. Via long, stringy nerve fibers that sprout from their surfaces, an electrical signal zaps from your fingertip to cells in the cortex of your brain in a fraction of a second. You perceive pain there, and at the speed of thought your brain fires a message back to groups of muscle cells in your arm, telling them to twitch in a coordinated fashion. The muscle contracts. Your arm recoils. All of this happens within a heartbeat.

The cut will have riven cells from one another in the walls of a blood vessel, a key event in kick-starting the healing process.

Opening a capillary causes blood to flood the wound. The scarlet of blood is hemoglobin, the protein molecule that ferries oxygen around your body, and it's packed into the concave disks of red blood cells shaped like a round mint half-sucked on both sides. Red cells account for just under half of the eleven pounds of blood that the average person carries. Most of the rest is made up of plasma, which is mostly water. But in that plasma, comprising less than 1 percent of blood, are cells that are utterly critical in repairing your wound. These are white blood cells, and their job is to find, fight, and thwart any opportunistic invaders such as bacteria, which immediately start trying to sneak into the body in order to flourish, but in so doing cause you infection.

Meanwhile, the tip of the nerve cell that triggered that pain sends out a signal in the blood that attracts platelet cells. These are the body's rapid response units, which clump together to form a clot to prevent further blood loss. They also act as emergency beacons, sending out signals to summon dozens of other "workers"—cells and proteins that protect the wound and initiate the process of rebuilding. Muscular cells in the artery walls spasm in synchronized contraction. Your finger throbs. This twitching restricts blood flow and loss, and helps keep immune cells where the action is. The formation of a clot prevents blood loss and hemorrhage, and marks the first phase of wound healing. Now that the barrier between the inside of your body and the rest of the world is reestablished, the cleanup and restoration can proceed.

After an hour, neutrophils comprise the majority of the cells attending the paper cut. These cells carry detectors on their membranes that pick up the chemical emergency signals pulsing out from ground zero, and move in the direction of the strongest of them. On arrival, neutrophils act as specialist cleaners, enveloping bacteria and vacuuming up debris and detritus, finally killing themselves when their task is complete.

Over the next twenty-four hours, another regiment of cells files into the site; each cell matures into the giant Pac-Man of the immune system, the macrophage (which translates into "big eater" in Greek). They chomp up the neutrophil carcasses and any other potentially damaging remains they find.

Crucially, the cut itself isn't simply stuck back together; otherwise we would lose the sensitivity that was there before the injury. Nor is it simply a case of plugging the gap with new skin cells; otherwise we would be lumpy and malformed. Our bodies strive to make repairs as invisible as possible, and to restore the body to its preinjury state. The cut needs to be patched up with new flesh, which is a complex collaboration of cells. And that means the birth of tissue.

As with any reconstruction, the foundations must be laid first. Building-block cells called fibroblasts flood the site of the cut over the next couple of days, reproducing themselves and migrating across the surface of the wound from all directions, extending ruffle-like feelers called pseudopodia—"fake feet"—as they shuffle along. The march stops when they meet in the middle, forming a layer of foundation for the full reconstruction, and they begin to turn themselves into parts of new blood vessels and new skin tissue. Locked down in position, they produce and ooze collagen to lay down a sort of matrix or scaffold for the rest of the reconstruction project.

Skin, of course, is made of cells, but not just one type. Our skin cells grow, layer by layer, from within, with dead cells sloughing off on the outside in a process of continual renewal. Embedded within this matrix are also a host of other cells, including hair follicles, sweat and oil glands, and the blood vessels that feed the flesh with oxygen and nutrients. All of these types of cells have to be rebuilt into the repair tissue.

A month after you opened this book, the cut is effectively healed.

But long after you have forgotten all about it, the cells of your body will continue work on the wound for months, maybe a year, remodeling the site to restore it as well as they can and minimizing any scar. The redness fades as the temporary blood vessels that extended to fuel the repair process retreat, and the temporary collagen matrix that acted as a scaffold for the rebuilding is replaced with a more permanent version. A new piece of living tissue, a new piece of you, has been formed in an act of minor but necessary repair.

The entirety of this reconstruction project has been brought about by thousands of cells working together and producing thousands of new, highly specialized cells that make up tissue: epidermis, glands, veins, and arteries. The fact that we, and all life, can create new living tissue out of cells is the grand idea that unites not only all living things, but every living thing that has ever existed.

There are more living cells on Earth than there are stars in the known universe. Take bacteria: my crude estimate is that there are something like five million trillion trillion (5,000,000,000,000,000, 000,000,000,000,000) bacterial cells on Earth. You are probably about fifty trillion human cells, give or take a few tens of trillion depending on your size. On top of all that we are all walking petri dishes. Healthy humans are born sterile, a clean slate of only our own cells. But in a typical adult the total human cell count is outnumbered ten-to-one by the nonhuman cells that we carry in our guts, on our skin, and every surface inside and out, more often than not in the form of bacteria. But there are also bacteria's cousins: archaea, as well as larger and more complex hitchhikers such as yeast cells, protists, and the occasional parasitic worm. Right now you are probably carrying more than a thousand alien species, most of whom you don't even notice, many of whom you couldn't live without. Mostly, these passengers are a fraction of the size of human cells, so by mass we're still mostly human. At a cellular level, though, we are mostly other things.

These numbers are so large as to be bewildering. But they do reflect that our planet is flooded with life—and that life is only carried by cells. Though the reconstruction project of your healing paper cut is awesomely complex, at the same time it is utterly mundane: a cut, a wince, and within a few days, new skin to patch it up. Similar processes occur billions of times every minute all over our planet. Not just paper cuts and wound healing but also the fundamental processes of all life take place through cells.

As we shall see, every single one of these cells, including the new skin cells on your finger, was born when an existing cell divided in two. And because of that simple fact, the cells in your cut have an ancestry dating back four billion years to the one exception to this rule—the very first cell. When new cells are born to patch up that cut, they are the latest in a chain that leads back, perfectly unbroken, to the very beginning of life on Earth.

How do we know this? What is a cell? How are cells able to perform this or indeed any process? How a cut heals may seem trivial, a biological act, albeit one of sophisticated orchestration, that is performed without thought. But in asking how such a refined process came to be, we will arrive ultimately at the same questions and answers that address the nature, and the origin, of life.

This is a book about those questions and their possible answers: the origins of our lives, the origin of all life, and the origins of new life. And at the heart of them all is our modern understanding of the cell.

Where life comes from is one of a small handful of the most fundamental questions we can ask. It is a question that has preoccupied humanity throughout its existence. Every culture and every religion has a creation myth: from the ancient Egyptians, who had a god sneezing, spitting, and masturbating to create the world, to the comparatively restrained Christian story in the book of Genesis, where life was created ex nihilo—out of nothing—and humans out of mud.

The true story began to reveal itself in a period of less than a hundred years between the middle of the nineteenth and twentieth centuries, with the emergence of the three great ideas in biology. As we will find out, cell theory, Darwin's theory of evolution by natural selection, and the discovery of the structure of DNA combine neatly to describe how life works. But they also bring us to the brink of cracking the big question itself: how life began.

Key elements of that narrative are elusive, and contemporary scientists are hard at work on filling in the gaps. With more fervor than at any time before, a model of genesis is emerging, using the weight of modern biology to reconstruct the deep past. It is only very recently, with our solid understanding of the genes, proteins, and mechanics of these living chemical processes, that we can seriously question how they came to be in the first place. Modern biology has revealed complexity rather than simplicity, and has shown that intricate networks of chemical reactions drive reproduction, inheritance, sensation, movement, thought, and all of the things that life does. None of this happens for free: energy is required to fuel these actions. The bottom line is that without energy, you are dead. In order to work out how life began, we have to unpick all of these networks. It is here, in the microscopic and, indeed, atomic world of the cell, that we are finding the clues to understand these processes—the ones that keep you, your cells, and every cell alive, as they have for billions of years.

As in contemporary astronomy and physics, the great theories of biology are now being tested with groundbreaking experimentation. Our exploration of the universe led to the big bang theory of the origin of everything, and at the Large Hadron Collider on the Swiss-French border the biggest experiment ever undertaken has been running to re-create the universe in its most embryonic form, billionths of a second after conception. In performing this act, physicists have unveiled the Higgs Boson, a most elemental particle that featured at the beginning of time and has continuously since then. So, too, the best way scientists can understand the pathway to life on Earth is also to try to re-create it. In the next few years, for only the second time in four billion years, a living thing, probably something akin to a cell, will be born in the laboratory without coming from an existing cell.

This act of creation will be a result of the story of biology. The knowledge acquired and the techniques invented in the last two or three centuries—but mostly in the last sixty years—have allowed us to dig into the deep past, but have also led us to invent the future. These two fields are intimately interdependent, a tangled thicket that shows how curiosity enables us to find things out. As we have revealed an ever-greater understanding of the processes at the beginning of evolution, we've learned how to profoundly manipulate biology in the present, and vice versa: as we take cells apart and reassemble them synthetically, we learn more about the conditions in which the first cells arose. Just like a mechanic learns how to fix, build, and augment the engine of a car by tinkering with, disassembling, and reassembling its parts, biologists can play with cells once they deconstruct them. The second half of this book explores the modification of life by human agency—at how we are designing, engineering, and building new life-forms for a purpose. It began some thirty odd years ago with what became known as genetic engineering—already a technology that has radically changed farming, disease treatment, medical research, and our understanding of living things. And in the last few years, genetic engineering has evolved and spawned a new field—synthetic biology—with a new approach to living systems that renders them tools.

This is a golden age in biology, and feverish work is occurring all around the world to resolve some fundamental questions. Our journey to this revolutionary point in the earth's history is the story of biology itself. And that story, too, begins with cells.

PART I

The Origin of Life

CHAPTER 1

Begotten, Not Created

Life is made of cells. For that reason, the vast array of these microscopic, gloopy bags is beyond description. In one single species, us for example, there isn't an absolute number, but of the fifty trillion or so cells that an adult might have, there are hundreds of types: from astrocytes in the brain to the zymogenic cells of the stomach. Along with that variety comes an assortment of dimensions. The longest cells are neurons in the spine, which stretch all the way to your big toe. If size is important, then you should turn to sex. Among the largest cells in humans are the eggs, just about big enough to see with the naked eye. And the smallest cells are their counterparts, sperm. However, what men lack in size they make up for in numbers: the average adult male can produce ten billion sperm per month, whereas women carry a finite number of eggs, one released every month from alternating ovaries between puberty and menopause. Women are born with all of their eggs already in place—the one that made you was made while your mother was growing inside her mother's womb, meaning that your first cell began its life inside your grandmother. Other than eggs, almost all cells are invisible to the naked eye, and even with a microscope most look unremarkable: tiny, colorless blobs bound by a fractionally less colorless membrane, generally sitting in a fairly nondescript, dimly lit swill. In labs, when we freeze tissue and slice it into slivers less than one-hundredth of a millimeter thick on glass slides, the cells appear packed together in dense abstract patterns. Or we grow them in broth, where they can be seen free-floating like blurry stars in an off-white sky. We stain cells in hues of pink and purple, and more recently in fluorescent greens and reds, to help visualize their inner workings. But in a live body, most are opaque as jellyfish.

Each type of cell is a highly specialized member of a community, working in unison with others to build a fully functional organism. Every process of our lives is a result of those cells performing their jobs. As you read this sentence, the muscle cells around your eyeball contract and relax to control the movement of your eyes from left to right. If you glance above this page now and look at something in the distance, a ring of muscle cells achieves focus by stretching the clear cells in the lens. You move your eyes effortlessly, but this action requires intricate unconscious coordination. Photons of light pass through your lens and hit the cone and rod photoreceptor cells at the back of your eye, in your retina. There they are harvested and converted into electrical impulses that zoom through neurons, via the optic nerve, up to the brain for processing, perception, and, with luck, understanding. Each movement: every heartbeat, thought, and emotion you've ever had; every feeling of love or hatred, boredom, excitement, pain, frustration, or joy; every time you've been drunk and then hungover; every bruise, sneeze, itch, or snotty nose; every single thing you've ever heard, seen, smelled, or tasted is your cells communicating with one another and the rest of the universe.

Douglas Adams once suggested that Earth was not the most appropriate name for our planet, as most of the surface is not solid dirt and rock, but water. Yet if you really wanted to name our home world after a feature that truly differentiates it from the other thousand or so planets that we have found, it would be Cells. Earth, uniquely as far as we know, is bursting with life, and every living thing on our planet is made of cells. Bearing in mind that nine out of ten things that have ever lived on Earth are already extinct, the number of cells that have ever existed is utterly incalculable.

This is a very modern understanding. Biology is a young science, at most 350 years old, and only 150 in terms of a fuller, mature view with comprehensive and universal rules. Physics has an older pedigree. By the mid-seventeenth century, scientists had mapped areas of the cosmos with future-proof accuracy. Isaac Newton was drawing up a set of rules that explained why things move the way they do, and why we can stand on Earth and not float away. But what are now known as the life sciences were a long way behind. The reason for this is that the starting point for most scientific advances is to look at things and work out why they are the way they are. Unlike the stars and planets, no one had even seen—or at least identified— a cell before 1673.

At that time, science itself was forming. Gentlemen scientists such as Newton and Robert Hooke had formed the world's first scientific body, the Royal Society. But the man who first peered into the minuscule world of the cell at the birth of cell biology was not one of the esteemed, bewigged gentlemen of science. The unlikely beginning of the story of biology must be credited to a Dutch linen merchant named Antonie van Leeuwenhoek.

The business of making and selling cloth was inextricably linked to the development of better optical lenses, as merchants checked the density of fibers and therefore the quality of their fabric using magnifying glasses like a watchmaker's loupe. Van Leeuwenhoek was a skilled and meticulous lens grinder, working in Delft, the capital of Dutch drapery. He specialized in a technique that involved pulling apart a hot glass rod and squashing the ends back to form a ball, but was secretive about this process, as it had made him the greatest microscopist of his day. Van Leeuwenhoek's lenses were tiny fat drops not much bigger than a peppercorn, and he attached them to handheld contraptions nothing like current microscopes. His were rectangular copper plates, about one inch by two inches, with a hole at one end to hold the rotund glass-bead lens. On one side there was a silver spike to hold the specimen in front of the lens, held by a screw that could be turned for focusing. It was the fatness of Van Leeuwenhoek's lenses that gave them their superior magnification.

At least, that was his technological advantage. His other key attribute was insatiable curiosity. Van Leeuwenhoek simply liked looking at small things through his lenses. While I hope the paper cut described in the introduction is entirely imagined, Van Leeuwenhoek deliberately invoked exactly the same repair process out of unbridled curiosity. In a letter published in the Royal Society's official journal *Philosophical Transactions* in April 1673, Van Leeuwenhoek wrote, "I have divers times endeavored to see and to know, what parts Blood consists of; and at length I have observ'd taking some blood out of my own hand, that it consists of small round *globuls* [sic]." We think that he was looking at red blood cells, and this appears to be the very first recorded sighting of individual cells.[1]

As his microscopy skills improved he began to look at all manner of bodily samples and fluids. He went on to scrape the matter from between his teeth and observed the bacteria that cause plaque. At the tail end of the seventeenth century, Van Leeuwenhoek was becoming something of a celebrity for his exploration of a microscopic kingdom hidden in plain view. King William III of England and other dignitaries visited him to see what he had seen. One discovery, however, was kept private: his own semen, although he attested in his notes that the sample was acquired "not by sinfully defiling

---

1. I say "think" because he also describes seeing *globuls* in milk, and these are likely to have been suspended droplets of fat.

myself, but as a natural by-product of conjugal coitus." In this act, which it is perhaps best not to dwell on, he saw sperm for what they are: single cells. He also discovered cells in a bead of water from a local lake and saw what we now loosely call protists: single-celled creatures that include algae and self-powered swimmers.

Van Leeuwenhoek was the first person to definitively see individual red blood cells, sperm, bacteria, and free-living single-celled organisms. This last group he gave a cute name, animalcules, and in the 1670s he sent drawings of his discovery to the Royal Society in London. The fellows expressed skepticism, not least because when they asked Robert Hooke, their resident microscope expert, if he could see the same creatures in water from the Thames, he initially saw nothing.

Hooke's expertise in looking at tiny things was unparalleled, having published a stunning and popular volume a decade earlier, *Micrographia: or Some Physiological Descriptions of Minute Bodies made by Magnifying Glasses*. Rarely has a book been so accurately subtitled. It contains, as you might expect, annotated drawings of very small things. His microscope was a simple six-inch tube with two lenses, and a baseball-size crystal sphere to magnify the illuminating flame. Many of the images generated by this kit are now very familiar, including a giant foldout of a flea, and a terrifying close-up of the eyes of a hover fly, which looks staggeringly similar to contemporary photos taken with an almost unrecognizable descendent of Hooke's tools: the electron microscope. Samuel Pepys picked up a copy of *Micrographia* and noted in his diary that it was "the most ingenious book that ever I read in my life."

Quite right. But there's a significant irony contained within this magnificent tome. One of Hooke's detailed illustrations is of a longitudinal cross section of some bark from a cork tree. There in the meticulous picture are conjoined units making up the overall structure. In the text, Hooke uses the term *cell* to describe these units. In reality they are the dead walls that once housed cork tree cells. He chose the word as it derived from the Latin *cella*, meaning "cubicle," but noted that they were air filled, which helped explain cork's buoyancy. Hooke had seen the remnants of cells, named them cells, but did not for a moment imagine that his observations were of universal living units, nor the permanent legacy of the name.

That is how cells were discovered. But where did they come from and how were they formed? With curiosity and technology, Van Leeuwenhoek had pulled back the curtain to reveal an undiscovered kingdom, but progress was fundamentally hindered by a resolutely unenlightened ideology.

The Origin of Cells

The origin of the cells Van Leeuwenhoek glimpsed remained elusive. Obviously, people knew that sexual intercourse between a male and a female often resulted in a new life, but, perhaps because most sex goes unobserved, an entirely fanciful view of the origin of cells remained vexatiously persistent.

Spontaneous generation was the most popular origin-of-life story for thousands of years. The first substantive explanation of spontaneous generation comes from Aristotle, the unsung father of biology. In his book *Animalia* (*The History of Animals*), written in the mid–third century BCE, he describes the genesis of certain species:

> So with animals, some spring from parent animals according to their kind, whilst others grow spontaneously and not from kindred stock; and of these instances of spontaneous generation some come from putrefying earth or vegetable matter, as is the case with a number of

insects, while others are spontaneously generated in the inside of animals out of the secretions of their several organs.

Animalia is a fascinating book, probably the first great biology textbook. It's crammed with observations and conclusions about a huge range of species, some of which are dodgy.[2]

In spontaneous generation, Aristotle described an idea that persisted until the nineteenth century, during which time dozens of examples materialized. The Roman writer Vitruvius casually referred to spontaneous generation when advising architects in the first century BCE:

> [L]ibraries should be toward the east, for their purposes require the morning light: in libraries the books are in this aspect preserved from decay; those that are toward the south and west are injured by the worm and by the damp, which the moist winds generate and nourish, and spreading the damp, make the books moldy.

In the sixteenth century Ziegler of Strasbourg claimed that lemming hordes originated from storm clouds.[3] Indeed, following Van Leeuwenhoek's discoveries, all manner of speculation on the origin of cells and life was put forward, all involving spontaneous generation. In seventeenth-century Brussels, Jean Baptiste van Helmont—marked in history as a respected scientist, the father of the chemistry

2. For example, Aristotle declares that some fish are neither male nor female, and these are prone to spontaneous generation, unlike their sexed brethren. We now know that all fish are sexual beasts, to the extent that many, such as clownfish and wrasses, display hermaphroditism—they switch sex when their environments demand it.

3. The notion that lemmings are suicidal is equally fantastic, a myth derived from images of mass migration, possibly staged, in the 1958 Disney film *White Wilderness*.

of gasses—described an experiment in which he sealed a sweaty shirt in a vessel with some wheat and let it ferment in his dank castle basement for twenty-one days. Lo and behold, this concoction gave rise to mice.[4]

It's all too easy to mock the ignorance of the past, and we should be wary of it. Though spontaneous generation was a persistent idea, it is not a solid scientific concept. All the misplaced examples, most notably when describing the genesis of larger beasts, were born of incomplete or poor observations. There are few more elementary questions than that of the origin of life, either in terms of the origin of new life from existing life or the more fundamental absolute origin of life, ex nihilo.

It was almost two hundred years after Van Leeuwenhoek first saw cells before the concept of spontaneous generation was thrown out. The final dismissal bears all the hallmarks of good science: rigorous observation and a testable, predictive hypothesis. But it also comes with the stamp of pure drama: an international cast and a generous helping of money, fame, and betrayal.

The Birth of Cell Theory

The quality of microscopes improved steadily throughout the eighteenth and nineteenth centuries, and the popularity of the study of small things grew accordingly. The biggest advances came not from exploring the microscopic animal kingdom, but from looking at plants and algae. That different parts of plants were composed of cells was apparent in the first few decades of the nineteenth century, though the ubiquity of cells in all living things was not. Much of this

---

4. As every biologist who has worked with them knows, gestation for the common house mouse is indeed around three weeks, but, more important, mice are sneaky, small, and hungry: where there is grain, you'll find mice.

work was carried out in Germany, and the textbooks carried names describing a range of observed structures: Körnchen, Kügelchen, and Klümpchen (granules, vesicles, and blobs). But while the description of tissues was progressing, the root of its genesis was not. It was not until 1832 that the birth of cells was first described. A Belgian baron named Barthélemy Dumortier watched cells in an algae grow longer and longer until a partition wall appeared and one cell became two. Other scientists soon replicated the work and observed it in different algae and plants.

Even without a consistent model for reproduction, the innards of cells continued to be explored. In 1831, Robert Brown studied orchid cells.[5] Within them he observed "a single circular areola, generally somewhat more opaque than the membrane of the cell." He called it the nucleus, the name it still bears today, and we now know it as the central office for the genetic code in all complex organisms.[6]

As with Dumortier's observation of cell division, Brown's assumption was that the nucleus was not universal. Many thought division was an exceptional means of cell birth, and no such observation of fission had been seen in animal tissue. Plant cells are frequently considerably larger than their animal counterparts, and so the study of flesh trailed the examination of leaves. The nucleus had been observed in some animal tissues, particularly brain cells, but again, was not assumed to be present in all cells. Compounding the issue, the most common type of cell in humans, red blood cells, does not contain a nucleus: it is shed during their development.

---

5.   Robert Brown is most famous for Brownian motion, which describes the haphazard microscopic pathway that particles make in a gas or solution when bombarded by atoms or molecules.

6.   While Brown certainly named it, the prize for the first observation of the nucleus goes, yet again, to Antonie van Leeuwenhoek. In a letter to Robert Hooke dated 1682, he describes smaller bodies within the red blood cells of a fish. Admittedly, it's an incomplete description, and doesn't imply any of the subsequent importance of the nucleus. But an unnamed modern editor at the Royal Society, where these letters are kept, casually scrawled in the margin, "Discovery of the cellular nucleus."

The names most closely associated with cell theory in almost every textbook are Schwann and Schleiden. The story Theodor Schwann told was that there was a eureka moment at the birth of cell theory. Schwann and Matthias Schleiden met by chance at a dinner in 1837. Schwann was an anatomist, short and nerdy; sometimes he would not leave his home for days at a time while he examined bodily tissue. Schleiden was a troubled, sometimes suicidal botanist who had been influenced by Robert Brown's identification of the nucleus. Botany and animal biology were separate fields, far from the unification that evolution and genetics would ultimately supply. Over dinner they discussed their work on animal and plant tissue, respectively. The conversation must have become even more thrilling for the other guests as their discussion progressed to the nucleus, that smaller body in the middle of cells. Schwann and Schleiden both realized that the nucleus was the same in plant and animal cells. They rushed without delay to Schwann's laboratory to compare notes. From this point on, the idea that all living tissue was made of cells took hold.

Dynamic though this legend is, Schwann and Schleiden were in fact only partial contributors in the development of a model of life based on cells, and significantly wrong about one major part. Much of the work demonstrating the nucleus in plants and animals had come before them, and the universality of cells had been suggested by others before 1837. It was probably Schwann who first used the phrase "cell theory," but where he and Schleiden were most significantly off the mark concerned the origin of new cells. Both described the formation of new cells starting with the spontaneous appearance of a naked nucleus in the spaces between existing cells. According to them, that nucleus acted as a seed from which the new cell would emerge, like a growing crystal. It's not as far-fetched as heavenly lemmings, but otherwise it was still reminiscent of spontaneous generation.

Our understanding of the origin of new cells can be largely attributed to Robert Remak—a lost hero of biology, and a victim of politics and race. Remak was a Polish Jew who spent his adult life in Berlin. To get the university position he wanted and deserved, he would have had to betray his Orthodox Jewish roots and be baptized, something he never did. Through his undeniably good science he was eventually given a post of lecturer, followed by assistant professor at the University of Berlin, but this post came with neither salary nor lab. Compare Remak's trajectory with that of his contemporary cell biologist Rudolf Virchow. Born into a well-to-do Prussian family, Virchow was flamboyant and bombastic. Ultimately, he would be described as "the pope of medicine" and "the sole instance of a full-fledged physician-scientist-statesman in our time." He was six years Remak's junior, but they were appointed to the University of Berlin at the same time.

After careful observation, Remak rejected spontaneous generation in all forms, including that which Schwann and Schleiden were describing. For a decade, he studied all manner of animal matter, including muscle and red blood cells as they grew in frog and chicken embryos; he only saw cells splitting, with one pinching in the middle like a belt on a balloon until it became two.

Virchow followed Remak's work, and year by year edged closer to his way of thinking—that cells were only formed by cell division. In 1854, Virchow declared that "there was no life but through direct succession," and a year later translated this into a Latin motto: *Omnis cellula e cellula* ("All cells from cells"). He was popular and prominent, and expounded this idea wherever he could, like in his international best-selling textbook *Die Cellularpathologie*. But there was no mention of Remak in any of these writings. Virchow had done little of the work, but had adopted his colleague's toil without giving him any credit. Remak was livid and wrote to Virchow about the Latin summary:

[It] appears as your own without any mention of my name. That you make yourself ridiculous thereby in the eyes of the knowledgeable, since you have no evident embryological expertise, neither I nor anyone else can undo. If, however, you wish to avoid a public discussion of the matter, I would ask you to immediately acknowledge my contribution.

We sometimes forget that science is undertaken by people, with all their personalities in tow. The scientific method is designed to equilibrate for all those idiosyncrasies and usually succeeds. But giving credit where credit is due remains a perpetual problem.[7]

Nevertheless, Remak, and Virchow for all his sins, had nailed it. Life is cells, and cells only come from other cells. But like a zombie, spontaneous generation still shambled along, lurching up yet again in France in 1860. The man who finally killed it dead was Louis Pasteur. Not yet famous for the sterilization technique that bears his name, Pasteur was ambitious, young, and had twice been denied membership in the French Academy of Sciences.

An experiment by a prominent proponent of spontaneous generation had reignited belief in its existence. Félix-Archimède Pouchet wanted to show that mold would spring forth from hay, even if the hay, the air, and the water used were sterile. Pouchet boiled the ingredients and then cooled them down with liquid mercury. As if by magic, mold appeared on the hay. The Academy wished to sort this out once and for all, and offered up a prize of

---

7. For the record, Virchow was not a bad guy, even though this looks like the act of a backstabbing rat. Throughout his career he was tremendously politically active, fought social injustice, and drove civic reform in Germany and Prussia with great success. One legend has it that his liberal views annoyed the Prussian prime minister Otto von Bismarck enough that he challenged Virchow to a duel. As the challenged, Virchow had the right to choose weapons. He selected sausages: one cooked, the other loaded with roundworm. Bismarck, the sausage-fearing Iron Chancellor, withdrew.

2,500 francs to the first person to resolve the theory of spontaneous generation.

Pasteur spotted the flaw: the mercury had a layer of dust on its surface, which was bringing in the mold. So he designed the simplest experiment imaginable. His version of Pouchet's setup was to have two flasks containing a sterile but rich broth, one that would soon go cloudy if exposed to microbial life. One flask was left open while the other had an S-shaped curved neck protruding to one side. Pasteur figured that microbes carried on airborne dust particles would reach the broth in the first flask, but that the swan neck would not allow these contaminants into the broth in the second flask.

Within days the open flask was cloudy. But the swan-necked flask was perfectly clear, and remained so indefinitely. As a control, Pasteur snapped off the swan neck and observed the broth growing cloudy over the next few days. Pasteur claimed the cash and was duly elected to France's scientific elite.[8]

The ultimate fate of this incorrect yet persistent theory is best described in Pasteur's own words, spoken at a lecture in Paris in 1864: "Never again will the doctrine of Spontaneous Generation recover from the mortal blow struck by this simple experiment [. . .] those who think otherwise have been deluded by their poorly conducted experiments, full of errors they neither knew how to perceive, nor how to avoid."

These are harsh words, but true. Thousands of years of biological superstition were swept away by an essential feature of science: the experiment. And with that swan-necked flask, cell theory was completed. Like all great theories, it's a fusion of ideas, based on observation and affirmed by experimentation. It is also one of the great

---

8. To further prove his point that contaminants were in the air, he repeated this experiment in different locations: some in dusty rooms, others in the relative sterility of over two thousand feet in altitude on Mont Blanc. The results were consistent: clean air, no growth.

nodes of biology. The work of dozens of men and hundreds of years of investigation into the stuff that life is made of can be summarized in two sentences:

1. All life is made of cells.
2. Cells only arise by the division of other cells.

The implications of this idea are profound, as grand theories should be. It covers all life, a simple but comprehensive description of the innumerable inhabitants of the living earth. But, as we already know, there are trillions of different types of cells. In the case of red blood cells, for example, those found in humans are different enough even from those of our nearest primate cousins that they cannot be exchanged without serious consequences. By the time we start looking at species as distant from us as chickens, we find, as Remak did, that their red blood cells do contain a nucleus, unlike our own. The first component of this grand idea shows that the diversity of life on Earth is embedded in the magnificent range of cells. The second shows how that diversity came to be in the first place.

Change over Time

At about the same time that Schwann, Schleiden, Remak, and others were observing how cells behave, across the English Channel a youngish family man was busy reflecting on his overextended gap year. Charles Darwin was slowly and meticulously putting together an overwhelmingly compelling case that described how creatures evolve. Evolution, the idea that species are not immutable, was already being contemplated as a concept in the nineteenth century. But the process by which they change was unknown. Darwin had spent five years traveling thousands of miles on HMS *Beagle*, collecting specimens from the far side of the world. Upon returning he married his cousin Emma Wedgwood; they were both grandchildren of the pottery magnate Josiah Wedgwood. They settled in Down House in Kent, and, untroubled by financial pressures, Darwin began chiseling out a splendid idea. In 1859, after years of scientific and personal struggle, he published *On the Origin of Species*.[9] In it he sets out the second grand unifying theory of biology, one describing the process by which evolution takes place.[10]

Unlike the microscopists in Europe, Darwin was primarily concerned with whole animals—the macroscopic world. He observed when comparing individuals in any population that there is a natural range for any given physical characteristic. This variation is the means by which those individuals can have a competitive advantage over others. In an imaginary population of anteaters, one with a slightly longer tongue than his contemporaries may be able to root out more juicy termites and may be better fed and healthier as a result. This could have the effect of making him live longer or maybe making him a more attractive mate to a female anteater. Consequently, this anteater may have more baby anteaters, each potentially carrying that longer tongue. After a few generations of that success, because that tongue is a useful thing, long-tongued

---

9. Darwin was ultimately prompted to publish by another explorer-biologist, Alfred Russel Wallace, who came up with effectively the same idea and described it in writing to Darwin. A kind and generous gentleman, Darwin suggested they present the idea together, which they did in 1858.

10. As a result of Darwin's being a scrupulous note taker, writing down pretty much everything he saw and did, his important life has been documented in meticulous detail. There is an ongoing leviathan project to digitize everything he ever wrote, scribbled, jotted, and doodled, and publish it electronically at *The Complete Work of Charles Darwin Online*. There you can read about everything from his experimental bassoon playing to earthworms to the wooden slide he had built for the central staircase in Down House to keep his many children entertained. As a result of his coming up with pretty much the best idea anyone ever had, the body of literature about evolution is deservedly both gargantuan and wondrous.

anteaters may come to dominate the population and become the norm. Over successive generations the species will change. In contrast to earlier theories, Darwin observed that it was not characteristics acquired during a being's lifetime that were passed on to offspring. Years of obsessive measurements led him to establish the principle that for each and every trait—an anteater's tongue, hair color, and so on—it was the variation of that characteristic across the population that conferred advantage for an individual. Those traits would become more common in a population because the advantage they conferred would play out as greater reproductive success.

There are other significant selective forces, such as the complexities of sex, where males puff themselves up to win a female, or females exercise seemingly whimsical choice over males. But natural selection is the overarching force that has shaped the living world in which we live. It's a system of trial, error, and revision. Evolution is blind and has no direction. Species are not more or less evolved, nor are they higher or lower, as they were once and sometimes still are described. Through iteration, they are merely better adapted to survive in their environments. In an experiment that really should not take place, an orangutan wouldn't last two minutes in the boiling submarine waters of a hydrothermal vent, despite being sophisticated enough to use tools in the jungles of Borneo. Down in the hot sea of a vent, though, hundreds of species, including the six and a half foot long giant tubeworm and dozens of species of bacteria, are quite content to eke out their existence. Change is the norm, and adaptation is success.

In the 150 years since the *Origin of Species* was published, millions of scientists have poked and pulled at the theory of evolution; unpicked, tweaked, and yanked at it in every conceivable way. They have observed countless species from aardvarks (or anteaters) to zebras to study their behavior. They have created simulations of myriad populations first with mathematical models, later with computers, and have pushed and pressurized their artificial environments to see how they adapt over successive generations. They have bred and crossbred inestimable species to observe how inheritance works, to see where advantage lies in the next generation. They have let tanks of bacteria reproduce for decades and witnessed descent with modification in action. In the modern era, we have deciphered a host of genetic codes and seen exactly the differences in DNA that reflect one species becoming two, each finding a niche into which they are better suited. We've seen populations of bacteria adapt to the hostile action of antibiotics and, distressingly, emerge resistant. While the initial model that Darwin laid out has been modified and fleshed out, the "one big argument," as he described it, has survived intact throughout the necessary battering that an idea of this magnitude demands. This is why it is called the theory of evolution by natural selection. The colloquial meaning of the term *theory* as a hunch, or guess, or plain old stab in the dark, is woefully puny next to its scientific meaning. When scientists talk about a theory they are aiming at the top of the pile of ideas: a set of testable concepts that all point to and predict a description of reality that is so robust it is indistinguishable from fact.

Darwin drew up his masterpiece as the study of cells was beginning to emerge from the stagnant pond of spontaneous generation. But his description of evolution is not about the beginning of new life; it is, as is implicit in the title, about the origin of new species. These arise when an organism has acquired so many mutated traits that they are no longer capable of reproducing with what were once their kin. When we are taught natural selection, typically we refer to prominent visible traits—antlers, hair color, and, once in a while, anteaters' tongues. But we now know how biology works at a cellular level, and can transpose evolution by natural selection to the microscopic world of which Darwin was largely unaware. The fictional anteater's tongue is longer because, by the chance of variation within its population, that individual has more (or possibly larger) cells in that tongue, and the genes that generate this disparity of tissue will be passed on through sperm or egg to the next generation. Similarly, when you get a paper cut the reason your platelets form a plug and a wedge to help stop blood loss is effectively because creatures carrying cells in their blood that didn't perform that clotting function as effectively have been deselected by nature thousands of generations (and species) ago (probably by allowing the creature concerned to heal less efficiently, or possibly leak to death). Crucially, we now know that what is being selected is not the individual, nor the cell, but the carrier of the information that bears advantage. As with all biology, the information that confers clotting is held in DNA inside cells, the molecule that will play the central role throughout the whole of this story.

Cell theory and natural selection are reflections of the same truth: life is derived. It's incrementally and ultimately spectacularly modified, but, in essence, life is the adapted continuation of what came before.[11]

Public acceptance of evolution by natural selection has waxed and waned over time, and was disputed by scientists at least in its first fifty years or so. But now, at least among scientists and those who generally understand it, natural selection is the only valid explanation for the variety of life on Earth. While science by defini-tion expects to be corrected over time, it now seems extremely

---

11. Darwin's idea has really upset a lot of people over the years, despite being self-evidently true, demonstrably true, experimentally true, and on all counts robustly so. Opposition to his radical but thorough argument was immediate and forthright, but so was support. "How extremely stupid not to have thought of that," said Thomas Huxley, Darwin's most pugnacious contemporary defender. It's a flattering sentiment, but masks the fastidious detail and stack of work that Darwin put into his opus. Conversely, cell theory appears to have upset exactly no one. It was ob-served, refined, and then has simply remained true. The laws of natural selection, the laws of genetics, and the workings of DNA (which we will come to shortly) are taught, fittingly, as absolute cornerstones of biology. But the principles of cell theory are simply assumed, and assumed to be correct. It's an odd oversight, but I suppose it's one that we shouldn't complain about.

unlikely that Darwin's idea will be replaced wholesale. When you tie it together with cell theory, both ideas are compellingly fortified.

Although the concept of evolution—simply that organisms change over time—predated Darwin, in 1859 this idea, as well as that of natural selection, was new and truly revolutionary. They both rebutted the prevailing view across the majority of human history, that creatures were each created distinctly. Without the hard-fought work of Darwin and the closely observed work of the eighteenth- and nineteenth-century microscopists, the idea of multiple routes for life, separate origins not just for plants, animals, fungi, but for every single type of organism, might have seemed plausible. Even taking into account the innumerable variety of distinct cell types, each with highly specialized functions, separate or plural origins might seem reasonable.

Instead, thanks to Darwin and cell theory, we can link every organism on the grandest pedigree. As Darwin writes in the final paragraph of his masterpiece, "There is a grandeur in this view of life, with its several powers, having been originally breathed into a few forms or into one." These are some of the finest words set to paper, oft quoted, but some things are just worth repeating. Yet he did pose a question in those last five words: "into a few forms or into one." Which is it? What lies at the base of the tree of life? A single form, a cell, or several? The answer to this deep-rooted historical question lies not in the past, but in the molecular guts of every single living cell. By examining the mechanism by which cells pass on their characteristics and by which those characteristics mutate, we will arrive at an answer to the question of whether life has a single origin—and we will begin to see what it might have looked like when it first emerged.

CHAPTER 2

Into One

"Life is a whim of several billion cells to be you for a while."

Groucho Marx

L ook out the nearest window and try to count how many species you can see. As I sit here at my desk looking at the view
into the garden, I can count forty species of plant (at least six
of which I can name), a spider, a squirrel, and a suspiciously well-fed
wood pigeon. I know that there are millions of other beings nestling
in the soil, feeding on the leaves, hiding in the mortar between
bricks. I know that there are tiny parasitic insects hitching rides on
the hairs of the squirrel and in the feathers of the bird, and that
those bugs themselves are serving as a rich habitat for thousands of
bacteria. I know that on Earth bacteria outnumber and outweigh all
other living things by count and by mass, and there are billions in
every spadeful of loam. As undergraduates my classmates and I
grew cultures of bacteria taken from swabbing our teenage skin and
streaking them out in their multitudes on agar plates. You can look
down at the end of your nose and be confident that there is more life

in that view than in the rest of the known off-world universe. We have classified around two million living species, and we know that number is a massive underestimate, as new ones are discovered and named every day. Life is bewilderingly diverse. What is more confounding, though, is its conformity.

It doesn't take a great leap of faith to see that we are closely related to chimpanzees or gorillas. Only a designer setting out to deceive or fundamentally lacking in imagination or effort would create things so similar and pretend that they were not cousins. It is obvious that dogs in all their human-crafted forms are related closely to wolves, and science confirms it. It takes a little bit more analysis to show that dolphins are much more closely related to hippos than to tuna, but dig deeper and it becomes obvious that all three species have common ancestry visible in their spines, eyes, and the bone structure (among many other things) of their flippers, feet, and fins, respectively. Yet it's admittedly harder to see the commonality between the leaf of a sycamore and the fin of a cichlid. Or a western long-beaked echidna and the recently discovered Malaysian mushroom *Spongiforma squarepantsii*. Or *Turdus philomelos* and *Candida albicans*: one is a pretty songbird, the other a creamy fungus, but both are called "thrush."

This zoo is made of cells of many different sizes and shapes. And yet all cells are basically the same, just as all gasoline-powered cars are basically the same. Cars have a chassis, an engine, some wheels (usually four), a steering wheel, and so on. The details of the engine, the design, the tires, et cetera comprise the difference between a Porsche 911 and a Pinto. Fundamentally, however, they are both cars, derived via successive iterations from a common ancestor, one that bore an early version of the internal combustion engine built from metal and powered by fossil fuel. And so it is with cells. Under the hood there are mechanisms and parts that vary the performance of a cell depending on its job, as part of an organism or standing alone. Cells, in terms of their overall structure, are basically the same, but are highly specialized to build the complexities and adaptations of all life.

The starting point for those basics of biology was the result of cell scientists making forward strides in the mid-1800s. This was clearly a feverish time, as Darwin was also deducing evolution, and a few thousand miles east, an Austrian holy man was planting a garden that would invigorate biology forevermore. Gregor Mendel is invariably described as a monk, which, while absolutely true, shrouds the fact that his legacy is unequivocally as a scientific genius and world-changing experimentalist.[1] Around the time Darwin was writing his masterpiece, Mendel had been studying pea plants and was breeding them in the tens of thousands. As any scientist will tell you, large numbers make for good statistics. What Mendel found in impressively large numbers was that, when crossing variants of pea plants with one another, the outcomes in the offspring were entirely predictable. He showed, moreover, that the traits were inherited in a discrete way—that is, independently of the rest of the plant's characteristics. Rather than a blending of flower color, the progeny of a purple-flower pea plant and a white-flower pea plant were not pinkish, but resulted in a predictable number of white ones and purple ones. By repeating these crossbreeds until the patterns were clear, Mendel's pea experiments also resulted in his determining that characteristics were inherited equally, one from each parent, but that some of those characteristics were more equal than others. He bred tall plants with short ones, and their offspring were always tall, rather than an average of the two heights. When he crossed those

---

1. Mendel went on to become abbot, but history does not record how good he was at monking.

offspring together, three-quarters of their offspring were tall and one was short. In those proportions he had uncovered not only that characteristics were passed down individually, but also that some characteristics were dominant over others.

The story of Mendel and his peas is high-school biology. What he had discovered (though the name came much later) was the existence of genes—discrete units of inheritance.[2]

Having been largely ignored, at the beginning of the twentieth century Mendel's papers were rediscovered. What followed was observation beyond that which is visible to the naked eye. New technologies of the twentieth century meant that the scale of biology was reducing from the organism to the cell, to the molecular and atomic level, and with this zooming in came the birth of modern genetics.

"It has not escaped our notice . . ."

Between Mendel's death in 1884 and the 1950s, there were major successive advances in the study of genes. Mendel had established that inheritance occurred in discrete units. Italian marine biologists looked at the cells of sea urchins and observed chromosomes—neat structures inside the nucleus of all cells, which became visible, resembling tiny sausages, when the cells divided. They came in

2. Mendel performed his key experiments between 1856 and 1863, and in 1866 published them in a minor parochial journal, *Proceedings of the Natural History Society of Brünn*. Mendel scholars have determined that 115 copies of the paper were distributed; many writers have described how one of them made its way to Darwin himself, and how it was discovered in his library after his death in 1882. But oh, the heartbreak of it all! The pages of this manuscript were uncut. Darwin never even opened it. Imagine the fusion of these great ideas, the two great biological geniuses finally uniting their work into one all-encompassing law, the mode and the mechanism of inheritance. . . . Alas, this story is fully untrue, a "what if . . . ?" myth. There was no copy of Mendel's pea paper in Darwin's library, cut or uncut. Indeed, according to the custodians of his library, Darwin had no published work authored by Mendel in his vast collection at all. Where this charming fiction has come from is unknown.

specific numbers depending on the host, and these marine biologists discovered that altering the number of chromosomes resulted
in abominations in their offspring, or prevented reproduction altogether. In the 1920s, Thomas Hunt Morgan inbred fruit flies to show
that Mendel's units of inheritance were positioned very precisely on
these chromosomes. German researchers, meanwhile, had shown
that chromosomes were made of a molecule called DNA, whose
chemical composition was clearly different from the proteins that
made up much of the cells ingredients, as it contained phosphate.

In the 1940s, Oswald Avery, Colin MacLeod, and Maclyn Mc-
Carty demonstrated it was DNA that conferred characteristics and
passed them on by performing a neat resurrection trick in New
York, perversely a fatal one. They were following the work of Fred
Griffiths, a British medical officer who more than a decade earlier
had noticed that, during the course of pneumonia, virulent and benign bacteria were both present, but that the latter could acquire the
malign characteristics of the former, even if these malign bacteria
were themselves dead. He demonstrated this by boiling the lethal
bacteria, thereby killing them, and then adding the resultant broth
to the benign bugs, which then acquired the ability to kill. Avery
and his team repeated Griffiths' experiment, but to figure out how
it worked, they systematically eliminated each of the components of
the bacteria that were candidates for passing on this characteristic.
It was only when they destroyed the deadly bacteria's DNA that the
transformation failed to occur. It seemed that DNA, not proteins or
anything else in the cell's milieu, was the essential genetic material,
the stuff that bestowed characteristics and passed them on.

DNA was clearly a—possibly *the*—key component of inheritance. But it was unclear how this could work. The answer lay in its
construction. At King's College London in 1952, a group of researchers including Maurice Wilkins, Rosalind Franklin, and Ray-
mond Gosling were investigating DNA using their expertise in

producing photographic representations of three-dimensional molecular structures. X-ray diffraction, a standard technique for establishing three-dimensional models of complex molecules, was brought to London by Wilkins from the Manhattan Project, which created the first atomic bombs. The principle is similar to the kind of silhouette portraits that were fashionable in the eighteenth and nineteenth centuries: shine a beam at your subject and capture the light and dark that is projected past it. The human subject in portraiture is a solid to the visible light that this technique uses, but in the molecular version the X-rays penetrate the molecule under scrutiny and create signature shadows behind it, regular but cryptic swirls cast onto the photographic plate. Mathematical deduction is required to figure out the arrangement of atoms that could produce such a pattern, but the effect is the same: a unique portrait of a molecule otherwise too small to see. Franklin was particularly skilled at this technique, and of the many photographs that Franklin, along with Gosling, developed while performing this arduous method, Photo 51 became the key to one of the great achievements in human history.

The Cambridge scientists Francis Crick and James Watson acquired the photograph. Science always builds on the work of others, but it was with their insight and genius that they deduced from Franklin and Gosling's photo that DNA took the form of a twisted ladder: the iconic double helix. On April 25, 1953, in a brief paper published in the scientific journal *Nature*, Crick and Watson showed that the rungs of this twisted ladder contained paired chemical letters—*A* for adenine, *T* for thymine, *C* for cytosine, and *G* for guanine. Each letter is bound to one vertical section of the ladder and pairs up with a corresponding letter on the other upright to form a rung. It is this pairing that makes the helix doubled, and the pairing is very precise: *A* always pairs with *T*; *C* always pairs with *G*. Crick and Watson concluded the paper with one of science's great understatements: "It has not escaped our notice that the specific pairing we have postulated immediately suggests a copying mechanism for the genetic material."

This is the first marvelous thing about DNA. If you split the double helix into its two component strands, you immediately have the information to replace the missing strand: where there is an *A*, the other strand should have a *T*, and where there is a *C*, the other strand needs a *G*. Therefore, DNA possesses an ability, inherent in its structure, to provide the instructions for its own replication. Thanks to Crick and Watson, building on the work of Wilkins, Gosling, and particularly Rosalind Franklin, we were given a molecule that could be copied and passed from generation to generation.[3]

The letter molecules, known as nucleobases or simply bases, bind to one another, holding the two vertical parts of the ladder together. There are millions of such pairings on each strand of DNA, something like six and a half feet of it in every cell that has a nucleus, though it is tightly wound up and then winds up on itself again around small lumps of proteins, like beads on a string. And this winds up on itself yet again, which forms a chromosome, like a thick tug-of-war rope.

The number and length of chromosomes varies hugely between all species, and that variation doesn't appear to relate to either the size or complexity of the host. We have twenty-three pairs, and bacteria tend to have just one, in a neat loop. But some species of carp have over a hundred, and this doesn't come anywhere near the number in some plant chromosomes, which can be in the thousands. The full collection of an organism's DNA, packaged into chromosomes, is called a genome. In humans, those twenty-three

---

3. Although we tend to think of DNA as the elegant double helix, in action inside cells it is very dynamic, constantly shuffling, twisting, splitting, folding, and remolding as it performs its myriad duties. When extracted in labs for experiment, DNA resembles pale, stringy mucus.

chromosomes, our genome, contain around three billion of those little letters, A, T, C, and G, which is sufficient to fill a two-hundred-thousand-page phone book, if that type of comparison is appealing. But every time a cell splits into two, from the first time your fertilized egg divides to the newborn skin cells in a paper cut, all the DNA in those cells copies itself. The new cell contains all the same DNA of its parent.[4]

Being able to replicate from generation to generation is a neat trick indeed. But alone, that would just fill the world with a pretty molecule. DNA's secret power is that this string of chemical letters, the bases, is a code that harbors information. That information is an instruction manual for all of the processes of life, including the very instructions required for the replication process itself. Understanding how DNA's code works will unlock not just how mutation and variation take place, drawing us closer to an answer to the question posed in Darwin's "into few forms or into one," but will give us essential clues as to how life was initially formed.

How DNA Works

All life is made by, or of, proteins. They form the structures and catalysts of biology, and the manufactories of bone, hair, and all the bits of a body that aren't actually made of protein. Naturally, this isn't limited to us, or even to mammals. Every leaf, strip of bark, reptilian scale, horn, fungus, feather, and flower is made of or by proteins. These workhorses of life are themselves made up of strings of smaller units called amino acids—a generic name for a

---

4. Biology is full of caveats and exceptions, and the exception here is a rather important one. During the process of sperm or egg production, a slightly different form of cell division occurs, called meiosis. This results in cells with half the total genetic material. There is a return to the full complement when sperm meets egg at conception.

potentially infinite number of molecular parts that qualify for this chemical moniker.

We now know that each gene—those discrete units of inheritance—in a genome is a piece of code made up of the specific pairings of DNA bases that encrypt the construction of a protein. However, only a few parts in many species' genomes are genes. The rest, in fact the overwhelming majority of DNA in humans, comprises notes, instructions, scaffolding, and even insertions from interloping viruses. Some are the remnants of genes from our ancestors whose function has been lost in us, but whose ghosts remain in our genome, free from the pressure of natural selection to slowly rust. Having established in 1953 that DNA was a corkscrewed ladder, and that it had the twin powers of replication and coded meaning, the biggest challenge in biology after Crick and Watson's great leap forward was to decode DNA. This meant first identifying which bits of DNA were genes and which were not.

Imagine that every sentence in this book were a single gene: the human genome would be a book forty times longer than this one, filled with random text but with my sentences distributed randomly throughout it. How would you identify which of them were the relevant sentences? Language is studded with punctuation to complement the letters and add composite meaning above the words themselves. Indeeditisprettydifficultforustoextractmeaningfroma sentencethathasnospacesorpunctuationinit.

Fortunately for science, and necessarily for the cell, DNA is no different. Before scientists could work out what the words and sentences of DNA mean, the first challenge was to work out where the spaces were—where a gene begins and ends—and this meant working out its punctuation.

By the 1960s, scientists knew that life was built of or by proteins, that proteins were built from amino acids, and that DNA was the hereditary matter that coded the proteins. The big gap was getting from one to the other, from DNA code to protein. By experimentally inserting a molecule in between two letters of functioning DNA, Crick and another future Nobel Prize laureate named Sydney Brenner intentionally disrupted its code and prevented a protein from being produced. These disruptions are known as frameshift mutations, like a film projector whose shutter speed is wrong, so you see half of one frame and half of another. By interrupting DNA with inserts equivalent to one, two, or four bases, the same effect was observed—a disrupted protein. But with an insertion of three, a protein product was still viable. Crick and Brenner deduced from this that the code of DNA works in sequences of three bases. This pattern is known as a reading frame: it places spaces in a length of DNA, exposing the meaningful triplets of letters.

Starting from there, an American and a German scientist cracked the first coded message in DNA in 1961 using a process of elimination. Instead of trying to work out how a naturally occurring sequence of DNA translated into a protein, Marshall Nirenberg and J. Heinrich Matthaei strung together a length of genetic code that only consisted of the base thymine (the letter T). They inserted this into the mechanics of a working cell and provided it with a ready supply of amino acids from which a protein might be built. They knew that there are only twenty amino acids from which all life is built, so in each of twenty different test tubes containing all of them, a single different amino acid within the mixture was tagged with a radioactive label. This meant that if a protein that resulted from their doctored DNA buzzed with radiation, they would be able to identify which amino acid was coded by a triplet of thymine bases. When they extracted the proteins from their various test tubes, the result was nineteen duds and one that set the Geiger counters buzzing. They discovered that a genetic code consisting exclusively of the letter T resulted in a protein consisting exclusively of the amino acid phenylanine.

And so the nature of the irrepressible code was known. Over the next few years, by varying the template, the unique triplets encoding the other nineteen amino acids were discovered, until by the end of the 1960s we had a complete readout of how DNA encodes proteins.

Here is a small section of DNA, part of a gene:

cctgggaccaacttcgcgaagcgggaagcccggcgg

Here is the same sequence broken into the triplet reading frame, as the cell reads it:

cct ggg acc aac ttc gcg aag cgg gaa gcc cgg cgg

And here it is again with each amino acid (here written in their abbreviated form) alongside each codon:

cct ggg acc aac ttc gcg aag cgg gaa gcc cgg cgg
Pro Gly Thr Asn Phe Ala Lys Pro Glu Ala Arg Arg

That string of amino acids forms part of a protein.

So how could scientists tell where a gene begins and ends? There is also punctuation in the language of DNA. In the continuous run of *A*s, *T*s, *G*s, and *C*s, the cell knows where a gene begins because without exception, they all start with the letters *ATG,* the so-called start codon, like a capital letter at the beginning of a sentence. Similarly, all genes end with a period, a stop codon, of which there are three: *TGA, TAG,* and *TAA.* A reading frame for an entire gene always begins with *ATG* and ends with one of these three stop codons.

Proteins are, then, long strings of amino acids as decreed by the DNA that encodes them. They perform their functions by folding up into three-dimensional shapes; the grooves, holes, clamps, and pockets in their folded shapes give them all manner of abilities.[5] Proteins also team up to gain new purposes. For example, hemoglobin, which carries oxygen around your body in red blood cells, is made up of four proteins, together carrying a single atom of iron. The astonishing properties of spider silk are the result of the sophisticated complex of different proteins of which it is made, some of them neatly folded, others overlapping to create a high-tensile strength comparable with that of steel. Some proteins are enzymes, which catalyze bodily reactions, the metabolism in cells that keeps us alive. Others are sensory, like the ones embedded in the rods and cones of your retina, so specialized that they can detect a single photon of light and trigger the process of vision. All these properties are the result of proteins folding, connecting, and interacting with others in highly precise ways. Diverse though the actions of proteins may be, the code underlying them is the same in all.

This is the bedrock of biology: the translation of code into action. But what is the process by which this translation takes place? If the genome is a sort of central office containing the plans to the factory, the plans never actually leave the office, so the relevant pages have to be photocopied. In other words, there is an intermediary between the DNA and the site of the manufacture of the protein, and this envoy comes in the form of DNA's cousin, RNA.

As previously explained, DNA is a helix twisted from two struts of a ladder, the rungs linking each together. But RNA is a single strand, the rungs exposed. When a particular protein is required, the double helix of DNA splits in two to expose the relevant gene. A single strand of RNA is laid down on top of the exposed gene, and each letter of that gene is copied into the RNA in mirror form. This

---

5. That sequence, from the retinal gene Chx10 above, folds into the shape of a grip that fixes onto short, unique stretches of DNA and instructs the cell to activate this gene, rather like using a word in the index to seek out a particular sentence.

envoy is called, appropriately enough, messenger RNA, for it carries the message from the genome to the site of protein construction.

Later on we will see that RNA plays a far more central role in the origin of life than its mere messenger status suggests. However, we have now arrived at the first and biggest clue to the question posed by the phrase "few forms or into one." The system described above is truly universal. There are no life-forms we know of that do not employ and entirely depend upon it: DNA, made of four letters, translates into proteins, made of twenty amino acids. It is known as the central dogma: DNA makes RNA makes protein. The fact that all known life is utterly dependent on this system makes it seem almost inconceivable that it is not related by a single, common origin. Certainly, working out how such a system might come about will be essential if we're to understand how life as we know it came to be.[6]

The Right Hand of Life

If the shared code and tools of life are not enough to point unequivocally to a single origin of life, here's an oddity of biology that emphatically seals the deal. Hold your hands up with your palms toward you. They are reflections of each other. If you slide one in front of the other, palms still facing toward you, one cannot obscure the other as your thumbs stick out. This, of course, is the basis of the glove industry: a left glove won't fit on a right hand. And so it is with

---

6. In a world that began to reject dogma during the Enlightenment, it is perhaps a pity that this pivotal process for all living things should have been named thusly. It was Francis Crick who called it dogma, and years later he expressed regret that his intended meaning with this word was perhaps not the same as everyone else's. Over the years, as with all rules and laws in science, it has been refined and modified, so although it remains utterly true, it's not quite as dogmatic as its moniker suggests. All the same, that's what it's called.

the molecules of life. In the same way that there are mirror forms of hand, so, too, are there mirror forms of certain molecules.

An atom is made up of a central nucleus, which is positively charged, surrounded by negatively charged electrons. These electrons form the bonds that hold individual atoms together in molecules and, like the poles on a magnet, these electrons repel one another. So they attempt to maximize the space between them. When atoms bond with other atoms to make a molecule, they will space those bonds as equally apart as they can, and so tend to adopt, wherever possible, symmetrical forms. An atom that has three available bonds, such as nitrogen, will tend to form a triangular molecule, with a bond at each of its corners. Carbon dioxide, meanwhile, is a straight molecule: the carbon atom has four available bonds, oxygen has two, and so the carbon atom is flanked by two oxygen atoms, each occupying a double bond. But if an atom naturally seeks to make four bonds, as carbon does, the farthest spacing of bonds is not in two dimensions but in three: a pyramid with a triangular base. The carbon atom sits in the center of this shape, with the four points equally spaced from one another. That's fine if all of the atoms on the corners are the same, or even if two or three of them are different from one another. But as soon as all four corners are occupied by a different atom, handedness becomes a possibility. In other words, exactly the same atoms in the same conformation can be arranged as mirror forms. In chemistry, this is called chirality, from the ancient Greek word meaning "hand."

We sometimes describe life on Earth as being carbon based, by which we mean that all DNA and proteins are built from frames featuring carbon atoms. And carbon, as we have seen, can form "handed" molecules, mirror images of one another, so it might be reasonable to suppose that we would find a mixture of right- and left-handed carbon-based molecules in life. What is remarkable is that, wherever life is concerned, the amino acids that make up proteins are left-handed.[7]

It was the killer of that persistently unfruitful idea of spontaneous generation, Louis Pasteur, who first identified chemical handedness. One of the ways we can spot the chiral nature of a molecule is by simply shining light through it.

Imagine tying a Slinky to a wall and then waving it in all directions. The waves would be up and down as well as side to side. If you then fed the Slinky through a vertical slit and did the same, only vertical waves would pass through. Certain molecules have exactly this filtering effect on light. They can polarize light into a single plane despite its natural tendency to wave in all directions. Louis Pasteur noticed that shining a beam through a solution of a simple molecule called tartaric acid that had been purified from wine finings caused the light to be polarized.[8]

Yet he also noticed that tartaric acid that had been made in the lab did not have this property, despite being chemically identical. The reason is this: the molecules in tartaric acid are based on carbon, and when synthesizing the molecule in the lab, both the left- and right-handed version are made in equal measure, the random chance of flipping a coin. This means that any polarizing effect of one version is canceled out by the opposite action of the mirror molecule. But the tartaric acid found in finings was naturally synthesized in yeast cells as part of the wine-making process. As a result,

---

7. The naming conventions that determine whether a chiral molecule is left or right are murky and not very helpful. In general, naturally occurring amino acids are referred to as L-amino acids, and they make left-handed proteins.

8. Pasteur was French, and therefore, I imagine, was probably jolly interested in wine. In fact, he developed the technique that bears his name—pasteurization—to sterilize not just milk but wine, too.

they are all right-handed. These partisan molecules would have allowed only one plane of light through.

This minor diversion into the realms of wine, chemical bonds, and spatial thinking is absolutely essential, because for reasons we don't fully understand, proteins only use left-handed amino acids. It is a mystery why all twenty amino acids that all proteins are made of are left-handed. Around four-fifths of humans are right-handed, so the equivalent with people would be as if southpaws simply did not exist.

Pasteur's revelation that the nonbiological production of handed molecules resulted in left- and right-handed versions in equal measure foreshadowed one of the most dreadful medical tragedies in history. The drug thalidomide is a mild but effective tranquilizer and painkiller. Its efficacy across a range of conditions from insomnia to headaches earned it the status of a "wonder drug." But it was also an effective treatment for preventing vomiting, and so was offered to pregnant women suffering from morning sickness. Between 1957, when it was introduced, and 1962, when it was globally withdrawn, more than ten thousand babies were born with severe birth defects including stunted limbs and other abnormalities as a result of being exposed to thalidomide in the womb. Thalidomide is a chiral molecule, and it was determined that, of the two mirror versions, only one had the mutating effect. As in Pasteur's tartaric acid, its production in pharmaceutical factories did not account for the two mirror versions, and the drug was sold with both hands in equal measure. We now know that either version of thalidomide has the confounding ability to flip to its reflection once in the body, so even in its harmless version, it would potentially be harmful to the developing embryo. It's still prescribed in some countries as an effective treatment for leprosy and other illnesses. But thalidomide's use by pregnant women is necessarily and strictly forbidden.

Handedness is a biological phenomenon, with proteins being

almost exclusively on the left. The origin of this preference is not known. It might be as simple as a chance event, but one that stuck. The uniformity points toward the development of a system with a single point of origin. If life had evolved twice, the likelihood would be that we would see both left- and right-handed proteins.

DNA has a similar but opposite steadfast bias: it is always right-handed. Point the index finger on your right hand and draw an imaginary clockwise circle in the air. At the same time move your hand away from your body. That is the turn of the double helix in its most common form, like a typical wood screw.[9] Even though that mirror-image double helix could perfectly well exist, it just doesn't. Life, yet again betraying its unique origin, only uses right-handed DNA.

The fact that a mirror world could exist but doesn't shows forcefully that life is of singular origin. If there ever was a choice between left and right, only one was selected, and the other discarded, never to feature again in the natural course of things. At the beginning of these biological mechanics, it might be as simple as a flipped coin, a chance event that was followed with absolute fidelity forever.

The Singularity

The origin of cells from parent cells and the origin of species via slowly changing genes in those cells both bear the hallmarks of a single origin. Those three aspects of biology—cells only from existing cells, DNA changing through imperfect copying, and modified

---

9. There is a website dedicated to outing depictions of DNA that spirals the wrong way, the Left Handed DNA Hall of Fame, for which I qualify alongside most of the major science journals and websites, not to mention newspapers, and many, many advertisements and films. There is one organization whose logo features the alien lefty screw, and that is the Astrobiology Society of Britain, composed of researchers whose work concerns the existence of life in the rest of the universe. Whether that was a design error or a deliberate play on the fact that they are primarily concerned with unearthly biology is unknown. As far as I'm concerned, it's quite clever.

descent of a species as a result—logically unveils a single line of ancestry that inevitably leads back to a single point in our deep, deep past.

In other words, any one of the new cells that were pulled from their parent as a result of the orchestrated disaster response to a paper cut bears a direct and noble lineage. Life is replete with extinction, but a new cell on your hand is one of the great survivors of the longest pedigree in history. In one generation its ancestry traces back to the fertilized egg from which every cell in your body was born. The egg and sperm that fused to form your unique combination of genes and DNA can trace a similar pathway back to their fertilized egg origin, and so on through their parents, grandparents, and every ancestor in your family tree—and, indeed, the history of our species.

Naturally, it doesn't nearly end there. These cells, in turn, trace their ancestry through the life span of our apelike ancestors. We don't know exactly who they were yet, but they existed in the bodies of individuals who first carved bones to make tools, started fires with flint, and stood upright as we, among apes, uniquely do.[10]

And it certainly doesn't end there, either. Before being in the loins of that upright ape, your cell lineage was carried through long-gone primates, maybe something like *Proconsul*, which resembled something between a chimpanzee and a macaque. Before them, it lay in monkeys and earlier through their furry, wet-nosed ancestors, more like modern lemurs. And as we follow them back, at some point, cells that were once the predecessors of your Pac-Man macrophages or sparking neurons existed in a shrewlike creature with fur and nipples. That critter, your ancestor, was present when an enormous meteor thundered into the Tropic of Cancer and called time on the reign of the dinosaurs sixty-five million years ago. Your

---

10.   It's not just the fossilized bones that tell us our past. Trace fossils are also crucial. The upright footprints set in soft ash 3.4 million years ago in Laetoli in Tanzania by *Australopithecus afarensis* are stunning evidence of our unique gait, but the truth is that we don't really know if these apelike people were our direct ancestors. The fossil record for human evolution is desperately uneven.

cell lineage witnessed the entire fall and rise of those beasts inside many forms of early mammals, such as the *Cynodontia*, which looked like a snarling rodent of unusual size. Before those first mammals appeared more than 220 million years ago, your cell's ancestor was a shelled egg of a reptile, hairless and cold-blooded, such as *Diadectes*, a six-and-a-half-foot-long brute that looked like a stocky, diet-shy crocodile.

To get to be in that egg, your cell survived iterations of beasts that had, among many other things, evolved cells that produce a thicker collagen framework in their breathing organs; with this innovation the primitive lungs could support themselves without the need for water. Your cell line may well have passed through a creature like the 375-million-year-old lungfish *Rhinodipterus*, which, unlike other fish at that time, had a muscular neck, which enabled it to raise its head above the water line. These innovations allowed it to breathe air rather than suck oxygen from the seas, and maybe opened up a whole new world of food. Further back your cell line was contained within a much more fishlike thing, with fins and gills. Before that it was in one of the first vertebrates, a swimming thing that looked a bit like an eel or a lamprey. And before that it was in a much wormier thing, similar to the living two-inch amphioxus: in one of those cells a massive copying error in its genome, a quadrupling of its entire DNA, resulted not in a fatality, but in a whole genetic platform on which vertebrates began to evolve. Before that it was in a spongy thing. And before that, just a clump of cells free-floating or lodged on a rock. And back and back and back ceaselessly into the past. The lineage of your cells has survived every disaster, catastrophe, meteorite, every extinction, ice age, and hungry predator—every event in this solar system for almost four billion years.

The vast majority of your cells, including all of the new ones in your cut, are terminal branches in this absolute pedigree, as their story will end with your life. The only ones that survive to give rise to cells in the next generation are sperm or egg. And of the trillions of cells that have worked for you throughout your life, only a handful will go on, the lucky few that will meet a sperm or egg and make a child. But the information within all of them will be carried on. Down that line has been transmitted the DNA that builds all of those cells together to make the most efficient way of ensuring the perpetual existence of your genes: an organism. Life is an astonishingly conservative system: DNA is the same in all species; the letters of the code are all the same; the encryption in the code is the same; even the orientation of the molecules is the same. What's true in bacteria is true in a blue whale. Only a system with a single root could display such conservation.

That pathway backward, retracing our steps as they become fainter and fainter over geologic time, can be applied to any creature alive today, or ever. The gaps become bigger, and it's almost always only hypothetical. Although we have a good overall understanding of the origin of species, to claim one species was the direct ancestor of another is often overstating what we can know. But the broad sweep of evolution is well understood, and any step backward through the past from any creature drives us to a single conceptual home. The branching tree of life ultimately becomes narrower as we reverse through time until we reach a single root.

We could trace an equivalent route backward for a cell taken from the boiling mud in an Icelandic hot spring, from the flower of a sweet pea, or from a button mushroom from the supermarket, and every time we would end up back in the same place. In every cell is a perfect unbroken chain that stretches inevitably back to the origin of life. That lineage irresistibly leads back to one single entity, which we call the Last Universal Common Ancestor, or Luca. Somewhere on the infant earth, Luca split in two. Since then, the thing that we struggle to define as life has passed uninterrupted from it to you, via a colossal series of iterations. Existence is bewilderingly tenacious.

Entities that we might be able to describe as living may have emerged several times, but life only enduringly survived once, and then continuously. We are confident of this because these other forms simply do not exist. At least they have not been discovered: there is a branch of speculation that proposes the rather futuristic-named shadow biosphere. This is the idea that there is a second (or more) undetected tree of life on Earth, with hallmarks different from the ones on the only tree of life we know of. But as it is, every life-form so far examined is based on cells, DNA, and Darwin. Discovery of a second tree of life here on Earth would give much-needed credence to the search for life on other planets, as it would double the number of known successful origin of life events. It would show that we are not a fluke. However, science is based on observable evidence. Therefore, the shadow biosphere, while sounding quite thrilling, is resolutely science fiction.

Pruning the Tree of Life

The next questions seem glaringly obvious. What was Luca, and where did it come from? We can reasonably assume that it had DNA as its genetic code, as all creatures after Luca do, and they are unlikely to have evolved that mechanism independently.

These ideas are largely based on what we share with other lives, whether it is the orientation of molecules, big physical characteristics such as five-digit limbs, or even simpler things such as having a head at one end and a tail at the other. Since the era of DNA began, and especially once the technology for reading genomes became much more accessible in the 1990s, studying evolution has been fortified by comparing similarities and differences in the precise letters of DNA.

Because we have shared ancestry, thousands of our genes are very similar in closely, or even distantly, related species.[11]

DNA acquires copying errors at a fairly steady rate, which means that we can compare DNA in any living species and figure out when they separated. We can compare genes and protein sequences from any two organisms and calculate how long they have been diverging. In this way, we can reconstruct history in exactly the same way as paleontologists do with fossil bones, by looking at similarities and differences, and assembling all those comparisons to show not just relatedness between two species, but when the schism occurred. This is called phylogenetics and has utterly confirmed Darwin's ideas about the branching tree of life.

However, biology is littered with exceptions of varying sizes, and when it comes to the tree of life, there is one colossal exception. Using phylogenetics, many scientists now argue persuasively that for the first billion years or so, life was not so much of a branching tree but a tangled bush.

The first life-forms, for the first couple of billion years following their split from Luca, were single cells. They were evolving, but not really transforming radically. In fact, despite reproducing thousands of times faster than most animals, for the first couple of billion years, life failed to make it past the stage of microbes. These two domains of life are archaea and bacteria—superficially similar things, both single-celled entities of roughly the same size. For a long time, archaea were similar enough to bacteria not to be recognized at all. But they are different enough to now be classed as very

---

11.   A gene called Pax6 is a beautiful example, and one on which I once worked. In humans, this gene encodes a protein whose role is to dictate that an area of the developing brain will mature in eyes. It does the same job in mice and zebrafish, whose version of Pax6 only differs by four out of a hundred amino acids, despite having a common ancestor around four hundred million years ago. The conservation is such that in the fruit fly beloved of genetics researchers, Pax6 also specifies where an eye is going to grow, even though it is a fly eye, not a mammalian one.

distinct from bacteria and, indeed, from everything else (and we will find that these differences are crucial in the theory of the origin of life). As a domain, they are a separate category at the highest rank of how we categorize life. The great leap forward occurred with the arrival of complex life. This branch of the tree, the third domain, is called eukaryotes, and includes everything that isn't in the first two, including you and me and yeast and snakes and algae and fungi, flowers, trees, and turnips. At some point, maybe some two billion years ago, complex life emerged when an extremely unlikely pairing occurred: an archaea swallowed a bacteria. Rather than death of one or both, mutual benefit was the result. The consumed ceased being a free-living entity and was permanently annexed into the guts of what would grow to become the third domain of life, the one that you are in.

This idea was first presented by the American biologist Lynn Margulis in 1966. It caused rousing controversy and was largely rejected, and she was considered to be heretical. With time, experiments, and evidence, her views were vindicated and are now the orthodoxy. The evidence comes in a number of forms, most simply that complex cells, from which we and all animals are created, contain small internal power units called mitochondria. The processes that happen inside these power stations are of fundamental importance to the origin of life, but we will come to that in due course. Mitochondria function as chemical engines that provide energy for the cell and, by extension, the organism. In simple terms, mitochondria resemble bacteria. They are roughly the same size as bacteria and, just like bacteria, have circular strings of DNA as their own genome, which are independent of the vast majority of a cell's genetic information, safely housed in the nucleus.

That engulfing of one life-form into the other was not a meal, but a hostile takeover. The engulfed would never be free again, but enabled previously impossible growth in its new host. It came with its own genome, with thousands of genes. Over time, most have been lost to natural selection, or migrated to the host's nucleus control center. But the mitochondria retains an independent set of genes to this day, almost all of which are devoted to maintaining the energy generation for cells. When this happened, charged with new energetic vigor, genomes could grow and form greater templates for evolution beyond single cells. Cells could evolve internal structures and compartments that increased specialization of cells. From there, the acquisition of coordinated communication between cells meant that an organism was not restricted to a single cell. Multicellularity followed, eventually enabling the evolution body plans for plants and animals, complex networks of cross-talking cells that interact with one another and the environment in harmony.

In backtracking through the tree of life in order to work our way back to Luca, these events provide a problem. The complication in estimating the timing and a description at the base of life comes from the fact that both bacteria and archaea do something else scientists did not expect. We pass genes down only from cell to daughter cell, from parent to child. Bacteria and archaea can swap genes, and therefore characteristics, with other individuals. Sometimes they don't even need to be of the same species to do this. This is called horizontal gene transfer (as opposed to vertical descent), and it is critical in confounding our attempts to understand the origin of life. The reason for this is because these cells don't always gain evolved functions by the typical process of descent via cell division.[12]

---

12. It is precisely this behavior by which so-called superbugs have become such an infectious problem in hospitals. Within a few years of the introduction of penicillin in the 1940s, a previously undetected strain of the bacteria *Staphylococcus aureus* was described that was resistant to penicillin. What this means is that in the presence of the otherwise lethal antibiotic, a random mutation appeared in one bacterium that rendered penicillin ineffective. As bacteria reproduce with alarming speed, soon this penicillin-immune microbe had spread. So we invented methicillin, another antibiotic that would still prove lethal to the penicillin-resistant ones. Guess what? Within a few years, methicillin-resistant *Staphylococcus aureus*

The evolution of language is a handy metaphor here. The words *bigamy, bicycle,* and *biscuit* have a common root. *Bis* in Latin means "twice," so you get twice married, ride on two wheels, and eat twice-cooked tasty snacks. But there are some words or phrases that are better expressed in other languages or even absent in the recipient, and so just get pinched. *Cul-de-sac* has a slightly different and more specific meaning in English than "dead end," so we have adopted it. The German word *Schadenfreude* has no English counterpart, but is now a splendid word in itself for ignobly enjoying your enemy's misfortune. Yet in Swedish, the word is *skadeglädje,* derived and adapted directly from *Schadenfreude* but bypassing any common root. It has passed horizontally and subsequently evolved independently.

What this horizontal gene swapping means is that our techniques for tracking back through the history of life using DNA as our guide may only work convincingly from the point where descent becomes vertical. The tree of life only begins to resemble a branching thing that deserves to be named a tree *after* the emergence of complex life. From that species on, when an archaea swallowed a bacteria, the overwhelming majority of inheritance was from parent to offspring: descent with modification. UCL biochemist Nick Lane calls this the genetic event horizon: comparing genes will take us all the way back to the point where it looks like a nice,

---

(MRSA) were found. Evolution has provided the mechanism for survival, and bacteria use it with tenacious vigor. Once a bacterium has randomly developed resistance to an antibiotic, it is more than happy to share it not just with its descendents but with its neighbors. They extend a thin bridge, a *pilus,* which brings two bacteria together, and through a tiny pore sections of DNA can be transferred from donor to recipient. This vastly speeds up the spread of an advantageous trait. It means that a population doesn't necessarily have to wait for a random resistant mutation to arise and to be passed from parent to offspring. It might be that the resistance is already in place, unused, and is deployed throughout the population as soon as they are exposed to the antibiotic. It's a familiar story: we try to control an organism, it evolves to survive. Humans waging war against bacteria is a bold gesture. Here, and elsewhere, the words of the great chemist Leslie Orgel resound: his so-called second law states that "evolution is cleverer than you are."

branching tree, but our vision grows blurred before that point. It just becomes far too messy to unpick the deepest past.

Therefore, while we can sensibly use DNA to infer that humans and chimps had a common ancestor around six or seven million years ago, and humans and the meager sea worm amphioxus had a common ancestor around five hundred million years ago, we cannot reliably use this to date Luca. The branches of the tree become tangled and scrambled before complex life, and tracing the changing patterns of DNA as species evolve is impossible. We are left, then, with very little genetic evidence of what Luca actually was at all.

The Creation of Luca

We can figure out some things about Luca, though. Calculations and logic predict that the last universal common ancestor had characteristics that are shared between archaea and bacteria, and that means genes, proteins, and the cellular mechanics largely similar to what we see today. That means we can use comparisons of these molecules to understand some things about Luca, even if we cannot apply an accurate date. A study by Douglas Theobald (Brandeis University, MA) in 2010 applied hard statistical analysis to the domains of bacteria, archaea, and complex life. He looked carefully at the construction of twenty-three proteins that are present in each of these domains with seemingly common descent, like words that sound and mean similar things in different languages. Based on the similarities of the sequence of amino acids that make up these proteins, Theobald calculated that the odds of their having arisen independently was 1 in 10^2,860 (i.e., a 1 with 2,860 zeroes after it).[13]

Another clue to Luca's singular origin concerns the most fundamental cellular machinery—the ribosome. It exists in all cells as a processing plant organized in minuscule blocks of molecules. Its role is universal, and as such, again supports the single-origin idea. The ribosome reads the genetic code and translates it into protein. The intricacies of this exquisite machine are explored later in the book (and particularly in the afterword, p. 240), but in essence its job is to read the genetic code (already transcribed into an RNA version) and translate each three-letter section into an amino acid. The ribosome strings together the amino acids in accordance with the messenger RNA, and a protein feeds out like a ticker tape.

We look to the ribosome as a useful indicator of relatedness because it is fundamental—without its manufactory we have no proteins, and none of the life we are aware of can exist. We can understand the idea that mitochondria in complex cells are derived from annexed bacteria because mitochondria have their own ribosomes, separate from their host cell, and these ribosomes are much more similar to bacterial ones than animal ones. We can compare the sequences of the genes that encode parts of the ribosome in all species, and trace backward the changes that have occurred over time to predict what Luca's ribosome looked like.

This is a fruitful endeavor, but the results are contested. For example, by looking at the particular sequence of ribosome parts across many species, you can reasonably infer the temperature at

---

13. Amazingly, this hasn't settled the matter. A Japanese team has published a response that says this analysis is not sufficient to reject the idea of multiple origins. That's not to say they or other critics are advocating a multiple-origin hypothesis, just that this particular study doesn't prove its own conclusions. Scientists can be a quarrelsome lot.

which that host organism thrives. Several parts of the ribosome are built from neatly folded RNA molecules, themselves made from letters of genetic code *A, C, G,* and *U.* As in the double helix of DNA, *C* pairs with *G* and *A* pairs with *U.* But *C* and *G* form a bond that is more stable at higher temperatures. Therefore, we can infer from the relative amount of *CG* bonds in ribosomes a preference for warmer conditions. This is borne out across many species, and the ribosome in which we see the highest *CG* content are from high-temperature extremophiles—in other words, organisms that thrive in heat. Where do we find such creatures? In many places, but the most impressive habitat is in and around hydrothermal vents under the sea. There, heated by a rip in the earth's surface, plumes of noxious chemicals that can boil the sea chug out. But dozens of species thrive there: bacteria, archaea, and even large, complex creatures such as the Pompeii worm, which can endure temperatures up to eighty degrees Celsius, more than 170 degrees Fahrenheit (not least because it wears an insulating fleece of hardy bacteria).

Some models of Luca's ribosomes suggest that the amount of *C*s and *G*s was disproportionately high in their component parts. This might suggest a hot home for the base of life. Certainly, the discovery of extremely heat-loving (aka hyperthermophilic) archaea and bacteria in such environments as the hot springs in Yellowstone Park or submarine thermal vents supports this idea, as these organisms mostly occupy spaces at the base of evolutionary trees, as much as we can reconstruct them.

But the simple truth is we don't—and possibly can't—know. Reconstructing the past using phylogenetics is a complicated art, with many confounding factors. Using the ever-changing DNA sequence, we can't see back to Luca because of the ability of bacteria and archaea to move genes sideways, not just from parent cell to daughter. That's not to say that Luca did not have specific characteristics that we can investigate and compare with those in living things. It's just that the idea of a single cell, a biological equivalent of Adam from the biblical Genesis, might be naive. If Luca was a cell, even in its most basic form, it still would have systems within it that are like their modern counterparts: DNA, RNA, proteins, ribosomes to make proteins, a cell membrane, and, crucially, a highly developed way of capturing energy—a metabolism. Otherwise, we would expect to see different and divergent mechanisms in bacteria and archaea. Not to denigrate this useful species; Luca's real use for science is as a proxy. If Luca is at the root of cellular life, it represents an all-dominant coagulation of what came before. Bill Martin, a brilliant and pugnacious origin-of-life biochemist we will meet properly later, says that the trouble with Luca is that, "like love, it means different things to different people."

It's been almost three and a half centuries since cells were bled, ejaculated, and fished out of their natural environment and viewed under a primitive microscope. Since then, we have picked them apart to the extent that we have earned almost full command of their faculties, acquired over four billion years. We see their commonality so clearly, a beautiful neatness where everything we discover in biology serves to refine and reinforce the truth of evolution. It's a wonderful state of affairs, and reflects the maturation of a science. The essential qualities of life are known and gel into a grand vision: life shares its tools, its processes, and its language. The security of a robust unifying theory of life as we know it allows us to investigate a much more difficult puzzle: where did Luca come from? As it happens, the best way to begin to understand the emergence of life on Earth is to take a hard look at where and when it happened. Therefore, we must start at the very beginning. It's a very good place to start.

CHAPTER 3

Hell on Earth

"Long is the way, and hard, that out of hell leads up to light."

John Milton, *Paradise Lost*

I f you want to construct a picture of the earth on which life first emerged, think about how we've named it. There are four geological eons spanning the earth's 4,540-million-year existence. The most recent three names reflect our planet's propensity for living things, all referring to stages of life. The second eon is called the Archean, which rather confusingly translates as "origins." The third eon is the Proterozoic, roughly translating from Greek as "earlier life"; the current eon, the Phanerozoic, started around 542 million years ago and means its name "visible life."

But the first eon, the period from the formation of the earth up to 3.8 billion years ago, is called the Hadean, derived from Hades, the ancient Greek version of hell.

Life does not merely inhabit this planet; it has shaped it and is part of it. Not just in the current era of man-made climate change, but all through life's history on Earth, life has affected the rocks below our feet and the sky above us. And necessarily, the origin of life is inseparable from the fury of the formation of the earth in the first place. A picture of the Hadean earth is crucial to understanding the wild natural laboratory in which life contrived to be born. Just as the formation of our home world is an event in space, we'll come to see how the emergence of life here is essentially a cosmic event.

The study of the early geology is all rock-hard science, but the evidence is often both literally and metaphorically thin on the ground. It requires geological detective work with clues dotted around the planet, and off the world, too. Geology gives us clues to how the earth formed from the bits and pieces of matter floating in space around the sun. Yet it also begins to describe the world that will evolve into the host of the only living things that we know of. While our lives are built on the stability of the earth, we also are keenly aware that our planet is sporadically violently active. The solid surface (including the seafloor) is made up of seven or eight leviathan continental plates and a collection of smaller ones. These all float on the slowly flowing but solid rock of the earth's mantle, itself encapsulating the molten core. The plates that form the crust are in constant flux, grinding and unhurriedly jiggling together. Some, such as the Pacific and North American plates, grind against each other, forcing up new land inch by inch. The subcontinent that is now India was once an island, and crunched into mainland Asia in a process that began around seventy million years ago, inching forward and crumpling up the land into the mountains of the Himalayas. These mountains will continue to grow at a rate of a few millimeters every year as the Indo-Australian plate continues to muscle its way into mainland Asia. Others plates are pulling the earth apart at the seams. The United States' colonial expansion continues westward every day, as the coast of Hawaii grows new land at a rate of many feet per year, with molten rock gurgling up above sea level and solidifying. Earthquakes shake the land and the seabed, dislodging mountainous blocks of water that become tsunamis, such as the one that wrecked the east coast of Japan in 2011. Still, these events are currently anomalies, though they remind us that our planet is alive not just with cellular life but also with slowly flowing rock. Yet for the most part our earth is reassuringly stable.

Not so in the past. The birth of planets is a process of summoning order from the chaos of the early solar system. Violence ensues. The sun, the star at the center of our planetary system, formed around 4.6 billion years ago as a colossal cloud of free-floating molecules collapsed under its own gravity and condensed into the huge nuclear fusion reactor that continues to heat the earth today. In the immediate aftermath, the sun sat in the center of a solar nebula, a flat disk of detritus left over by its formation but held there by its own gravity. Over the course of the next few million years, this matter, mostly dust and gas, began to stick together in clumps. At first these were the size of medium-size concert halls, but over hundreds of centuries, lumps collided and stuck together in a process called accretion. Closer to the hot sun, the temperature is higher, which makes it harder for gasses to condense and accrete. It's for this reason that the inner four planets of the solar system—Mercury, Venus, Earth, and Mars—are terrestrial, made of rock, whereas the outer four—Jupiter, Saturn, Uranus, and Neptune—are gaseous.[1]

It's impossible to describe a whole planet in simple terms, as we know very well how our home is not a unified land mass. Much of the surface of modern Earth is solid; more is ocean. Of the grounded parts, there are extremes of temperature and geography, snow, deserts, marshes, forests, plains, mountains, and so on. The rock beneath your feet is likely to be very different from the rock underneath a reader's feet on the other side of the planet. Similarly, it's virtually

---

1. Pluto, once the ninth, is no longer considered a fully fledged planet, as it is one of several similar-size bodies in that area of our solar system.

impossible to describe simply what Earth looked like in the Hadean. The evidence is vanishingly sparse, and as a result this period unhelpfully gets called the Cryptic era. Yet we can extract meaningful models in broad terms. We know that immediately after accretion, the planet was probably largely molten. Contrary to previous thinking, however, this period of molten Earth is now thought to have been short-lived. There are no surviving rocks from that era, so in the past we had assumed that there were no rocks. Yet the absence of evidence is not the same as evidence of absence.

Just as we look in rocks for the traces of recent life—the fossilized forms of dinosaurs, or even cells, for example—our planet's history before life is embedded in geology. The oldest dated matter on Earth comes not from rocks, but in the form of zircon crystals, an abundant mineral found all over the world but probably best known as a cheap substitute for diamonds in jewelry.

Zircons have two convenient characteristics. The first is that they can withstand metamorphosis—the brutal churning of rock over very long periods of time. The second is that the atoms of zircons are naturally arranged in a neat cubic structure. This molecular box can trap atoms of uranium inside them, as few as ten parts per million. A small proportion of uranium, like many elements, is radioactive, and over time will decompose into lead. Due to the precise nature of their crystal structure, when zircons form they can include uranium, but exclude lead atoms. Once imprisoned in this cage, radioactive uranium slowly mutates into lead over the course of millions of years, and because this decomposition happens at a fixed rate (what we call a half-life), the moment a zircon crystal forms, it sets a clock at zero. Lead found inside zircons must have begun its life as imprisoned uranium, so by quantifying it we can date the origin of that crystal with 99 percent accuracy. Out in Jack Hills in Western Australia, zircon crystals have been found that trapped their uranium 4.404 billion years ago.

Dating is not all we can tell from the constituents of cheap jewelry. We can also infer from those same crystals that their formation was part of a solidifying process, the development of a crust. This means that, although there are no rocks from that period, we can know that there was land in the Hadean. We can also determine what other ingredients were present. The Jack Hills zircons also harbor a particular type of radioactive oxygen whose presence looks like that found in modern crystals, which are formed as the earth's crust gets sucked down beneath the ocean floor. The presence of this oxygen suggests that from as little as a hundred million years after the initial molding of the earth, water was present. On Earth, there is no life without it, though this water was likely to have been extremely acidic.[2]

Therefore, the early Cryptic days on Earth might not have been quite the hellish inferno of endless seas of molten lava. Within a mere hundred million years, the earth had a solid surface and oceans. Sounds pleasant enough, but let's not paint such a picturesque portrait. The earlier theories of a molten Hadean Earth were based on a solid observation: we can't find any rocks from that era. If there was a rocky surface, what in hell's name happened to it?

It turned out that we were looking in the wrong place. In fact, we were looking on the wrong heavenly body altogether. To address the question of what happened to the early earth, we sent twelve men to the moon. The moon itself was born out of the most destructive impact that Earth has yet suffered. Somewhere between fifty and one hundred million years after the formation of our solar system, the earth had its worst day. It was struck by Theia—a terrifically and

2. The source of the earth's water remains controversial. The absence of an atmosphere and its position in space so close to the sun may have caused much water on Earth to evaporate. Yet perhaps the shape of the rocks that first accreted to form this planet held water in their crannies. Another idea that some scientists promote is that most of the water on Earth was delivered in one or several icy comets whose frozen payload melted on delivery.

inappropriately pretty name for such a harbinger of doom. Current theories suggest that Theia was a rock the size of Mars. It blasted enough matter from the embryonic earth into space that it reformed as our closest celestial neighbor: the moon. The impact was devastating, enough to rip the first atmosphere from the planet. Theia's glancing blow may be what shifted the earth's axis from vertical to its off-kilter stance of 23.5°. This lean is what causes the seasons, as when we are tilted away from the sun, each unit of solar radiation spreads out over a much larger area.

But it was what came after the formation of the moon that concerns us. It's the characteristic pock-marked lunar visage that gave us clues to the state of the Hadean earth. Between 1969 and 1972 the Apollo program of the U.S. National Aeronautics and Space Administration (NASA) landed six missions and twelve explorers on the moon, beginning with Neil Armstrong's famous first small step. Over those missions, astronauts collected around half a ton of rock and brought it back home for analysis. The last man to walk on the moon, Commander Gene Cernan in Apollo 17, is quoted as saying that "we went to explore the Moon, and in fact discovered the Earth." There is a great truth in that quotation, as it was in the subsequent analysis of lunar rocks that we were to discover the nature of the earth's formative years. Unlike the earth, the moon has no atmosphere or winds, and no shifting geology, so the craters formed by meteorite impacts are left undisturbed alongside the footprints of the Apollo pioneers. What that means is that we have a record of meteorite activity in the local solar system, footprints unscathed by the winds, seas, and tectonic grind of the earth. Geologists dated lunar rocks bearing the hallmarks of meteor strikes. These are called impact melt rocks; they all occurred in a precise window of time, between 4.1 and 3.8 billion years ago. We can deduce that this was a period of intense local meteoric activity, and by inference, that the earth also suffered this hellish pummeling from above. The young solar system was crowded with debris and leftovers from its birth, and for a period of three hundred million years, until the end of the Hadean, we got the full brunt of it. This period is called the Late Heavy Bombardment—so named because it was mercifully the last time Earth would suffer such a battering.

How heavy is heavy? Meteors fall from the skies all the time. Almost all, thank goodness, are tiny, burn out, then fade away in the atmosphere as shooting stars. Even big ones, such as the dramatic fireball that lit up the sky above Chelyabinsk, Russia, in February 2013, mostly disintegrate in the atmosphere. Occasionally, a big one hits, whereupon they become meteorites, such as one that fell on the small Australian town of Murchison in September 1969, just a few weeks after Apollo 11 returned Armstrong, Buzz Aldrin, and command module pilot Michael Collins to Earth. That one weighed more than two hundred pounds, and carried a payload of interest to this story, as we will find out later.

If you're prone to making a wish when you see a shooting star, why not hope that we don't get to witness anything near the size of the best-known meteorite. Sixty-five million years ago a five- or six-mile-wide rock smashed into an area of what we now call Chicxulub in Mexico. The crater is now hidden, mostly beneath the sea, but its 110-mile-wide shadow remains and was spotted by oil prospectors in the 1970s. In the ground and seabed, there is a broken but detectable ectopic circle of tiny glass beads that were forged from molten rock during the heat of impact. And from space we can see the same circle in minuscule gravity distortions only measurable from precision equipment in satellite orbit. There has not been an impact anywhere near that magnitude since then, and be thankful for that. The Chicxulub meteorite was the trigger that extinguished the reign of the dinosaurs and paved the way for small mammals to evolve, eventually into us. An impact of that order means that it's very likely that the meteorite instantaneously wiped out many millions of

creatures, with an expanding circumference of mile-high megatsu-
namis racing away from the impact site, leveling the land like a wave
crashing on a beach. With that there also would have been a fireball
hot enough to melt sand and rock into those telltale glass beads. But
the full impact of the meteorite would have taken thousands of
years, a dust cloud thrown up that blotted out the sun. The Chicxulub impact irreversibly changed the earth's system, wiping out once-
dominant life-forms. And yet compared to what was happening on
the infant planet, the place on which life began, Chicxulub was a
drop in the ocean.

Scientists have estimated that during the Late Heavy Bombardment something like fifteen astronomical rocks over one hundred
miles wide, twenty times the size of Chicxulub, bruised our world.
Of these maybe four of them were two hundred miles wide. For
three hundred million years, giant rocks rained down from the sky,
some as big as decent-size islands. The power of any one of the tens
of thousands of impacts during this time would make the most destructive nuclear bomb seem like a firecracker. Global environmental destruction would have occurred at least every few centuries.
Any potential surface habitat for living organisms would have been
destroyed over and over and over again. The relentless pounding the
planet suffered during the Late Heavy Bombardment was enough to
boil the oceans and vaporize the land.

And then it significantly calmed down. The meteoric blitz of the
Hadean ended around 3.8 billion years ago, leaving a frazzled earth,
still tempestuous and rough, but at least not barraged from the
skies. The sun was dimmer than today, probably less than three-
quarters of its current strength. Because of that, the earth cooled
quickly, and water from volcanoes and comets condensed into
oceans that covered the planet.

The precise point at which life began is unknown, and almost certainly unknowable. It may be that it began multiple times, maybe during the Hadean, but was wiped out all but once by the sterilizing spray of the Late Heavy Bombardment. One 2009 computer model by scientists in Colorado suggests that even had the Hadean eon sterilized the surface of the Earth, life could have survived at the bottom of the ocean.

The consensus (though not unchallenged) is that the first evidence for living matter dates to around 3.8 billion years ago, coinciding with the end of the Late Heavy Bombardment. These clues come in the form of that vitally important atom, carbon. Cells, defining life as we know it, are not visible in the fossil record at this age, as rocks older than 3.5 billion years tend to have undergone the harsh geological metamorphosis that irretrievably churns up any shadow of living structures. Therefore, we have to look for the chemical signatures of life trapped in rocks. In a formation on the west coast of Greenland, rocks have been found that contain the merest trace of a form of radioactive carbon that has no earthly reason to be there, unless it had been processed by a living organism.

We don't know what that life-form was: only by the presence of that carbon can we infer that an organism that had similar fundamental mechanisms to modern life existed all that time ago. Skip forward four hundred million years and the remnants of life are abundant and much less controversial.[3] The best of these comes in the form of stromatolites: foot-wide stone mushrooms that sprout from the shallow seas in Australia and other locations around the world. They are formed when floating mats of solar-powered

---

3. I say "less controversial," as some scientists contest that stromatolites could form from a nonbiological process—abiogenesis. Nevertheless, the consensus leans heavily toward these rocks being strong evidence for an Archean world thronging with microbial life.

bacteria trap tiny particles of grit in their slimy mucus, and over millennia this floating scum slowly settles into layered lumps of stone.

Ingredients

Yet that is hundreds of millions of years of evolution after the end of the Late Heavy Bombardment. What we see from then are scant pieces bearing evidence of living things in a colossal jigsaw. We have a picture of Earth, more settled than it had been for hundreds of millions of years, but still violent, with electrical storms, churning land masses, volcanoes chugging out gasses into the atmosphere, and tumultuous seas. This is a contemporary understanding of the early earth, and has helped us formulate experiments and hypotheses for the conditions in which life emerged. However, the first speculations about life's emergence predate our current ones by a century.

In 1871, Darwin wrote a letter to his friend Joseph Hooker contemplating the switch from inanimate chemistry to life. On the second page of this almost unreadable document[4] he considers not the origin of species, but the origin of life:

> It is often said that all the conditions for the first production of a living organism are now present, which could ever have been present. But if (and oh what a big if) we could conceive in some warm little pond with all sorts of ammonia and phosphoric salts,—light, heat, electricity &c. present, that a protein compound was chemically

---

4. Darwin's handwriting was extraordinarily sloppy, to say the least, and the original letter is barely legible.

formed, ready to undergo still more complex changes, at
the present day such matter wd be instantly devoured, or
absorbed, which would not have been the case before living creatures were formed.

With this famous "warm little pond" Darwin is prefiguring the
concept of primordial soup (*primeval,* meaning "original" or the
"earliest time," and *prebiotic,* meaning "before life," are also used,
largely interchangeably). He lists the ingredients of the soup there,
just like a recipe. Though ignorant of our modern picture of the
Archaean earth, Darwin had casually meandered into what would
become the dominant idea of the origin of life. He was not the only
one to take these speculative baby steps. One of his biggest champi-
ons was the German zoologist and polymath Ernst Haeckel, an early
proponent of the idea that biology and chemistry were continua-
tions on the same spectrum. In 1892, he proposed a process in
which "the origin of a most simple organic individual in an inor-
ganic formative fluid, that is, in a fluid which contains the fundamental substances for the composition of the organism dissolved in
simple and loose combinations." Chemists were already dabbling in
biological alchemy, not gold from base, but conjuring the molecules
of biology from chemistry. In 1828, a German scientist, Friedrich
Wöhler, synthesized urea, a key biological molecule and a component of urine, noting in his methods that he had done it "without the
use of kidneys, either man or dog." This contradicted the then-
popular concept of vitalism, that life was somehow fundamentally
different from nonlife. Wöhler had shown that the molecules of life
could be made synthetically.

The idea that the birthplace of the first life was a rich pond of
ingredients was formalized in the 1920s when a Russian, Aleksander
Oparin, and an Englishman, J. B. S. Haldane, both independently
wrote about the emergence of complex biological molecules and life

in conditions fueled by an oxygen-depleted atmosphere on the early earth. Haldane—a truly great scientist and a gifted science communicator who went on to become a central figure in the emergence of evolutionary biology in the twentieth century—is an important character in this scientific journey, as he first used the phrase "prebiotic soup," and the idea of soup as the stock of life has bubbled away ever since.

Soup had its greatest moment in 1953, a vintage year for science. Crick and Watson revealed the structure of DNA in April, certainly the scientific achievement of the twentieth century. But at exactly the same time, a young student was building on Haldane's and Oparin's ideas to put together another, similarly iconic experiment. Stanley Miller was a twenty-two-year-old chemist at the University of Chicago, and as part of his PhD he begged his supervisor, Nobel Prize laureate Harold Urey, to let him put together a rather fanciful experiment. He built a set of interconnected glass pipes on an electrified metal grid six and a half feet square. This kit now sits in a dimly lit room in the labs of one of Miller's former students, Jeffrey Bada, now an emeritus professor at the Scripps Institution of Oceanography in San Diego. It looks, not inappropriately, like a 1950s sci-fi experiment, with sparks and bubbling gasses and colored liquids. Miller filled the glass beakers with water, methane, hydrogen, and ammonia, in an attempt to emulate what was believed at the time to be the essential ingredients of the early earth. Moving on to the next stage set by Oparin and Haldane, Miller reasoned that the absence of oxygen in the atmosphere was key to the chemical maelstrom necessary to prompt the emergence of essential biological molecules. Miller put thousands of volts' worth of spark into that pipework, mocking the electrical storms and lightning that thundered from the turbulent Archaean sky.

Harold Urey had the good grace to let him pursue this experiment with the caveat that it would be shut down and he would have to move on to less implausible research if it bore no fruit within a few months. No such time was needed. Within days the mix turned pink, then coffee brown. Miller extracted the rich brew; his analysis of it found the presence of the amino acid glycine and a handful of the other biological amino acids essential for building proteins. He published his results in the journal *Science*, noting in the methods that the conditions were designed to emulate the primitive earth, not to optimize the production of amino acids. Just to consolidate that extraordinary result, there's a sweet coda to this experiment. In 2008, a year after Miller died, Jeffrey Bada rediscovered some of the original samples from this experiment tucked away in the back of a dusty drawer. He then subjected them to twenty-first-century analyses. Even in these fifty-year-old samples, precision inspection revealed not just the few that Miller had seen, but all twenty of the biological amino acids, and five others, too. It seems that under those conditions, the spontaneous production of essential biological ingredients was a trivially straightforward happenstance. Those simple units, it might be thought, would string together into proteins that all life depends on, and with their function the processes of life could begin. Miller had shown that in the tumult of the Archaean earth, the molecules that form the universal workers of living systems—proteins—were simply summoned into existence by the equivalent of a bolt of lightning.

The experiment was so appealing that Miller became an international celebrity. The press was amazed and excited; their reporting exaggerated the results to the point where some claimed that Miller had created life. Naturally, amino acids are not life, though they are essential for it, but their creation was big news. This experiment was seen as a stamp of approval for the idea of primordial soup, cementing its place as part of our culture, and the most tenacious idea in the origin of life. In a location somewhere on the earth, a wet surface or pool or floating pumice was exposed to the gasses of the

Archaean atmosphere and a bolt of lightning. It carries a sense of drama: the spark of life injected into a corpse, invigorated from the skies—the moment of creation.

But could it have been like this? As we have seen in earlier chapters, the mechanics of life are mesmerizingly complicated. A cell is a hive of densely packed activity receiving input from its environment (whether this environment is as part of an organism or as a single free-living cell). Inside the cell, there is code that encrypts proteins, and those proteins enact the functions of living: feeding, communicating with other cells, and the reproduction of the organism to perpetuate the genes it bears. To see the spontaneous emergence in a test tube of the molecules (or components of those molecules) that perform these vital acts certainly provides credence to the idea that the origin of life is neither mystical nor supernatural. At the beginning of the Archaean, simple chemical ingredients did make the transition from chemistry to biology, and Miller's iconic experiment follows in the direct scientific lineage of Darwin's warm little pond. In that charming speculation the recipe includes "ammonia and phosphoric salts,—light, heat, electricity &c.," all of which are plausible components as they relate to some of a typical cell's processes. Miller's experiment tested this idea, basing the ingredients on a better understanding of the conditions of the infant earth—ammonia, methane, hydrogen, water, and lightning. In the sixty years since, many experiments have further refined the recipe, or shown similar and more sophisticated spontaneous construction of biological molecules out of a soup of ingredients. It's an attractive idea, and one that has stuck. But there are fundamental outstanding questions that underlie these experiments. Can life emerge by cooking up a chemical soup? Is a spark what it takes to drive a chemical reaction into a biological one? To answer these questions, and to get to the bottom of the origin of life, we must ask a very simple question, one with a deeply elusive answer.

CHAPTER 4

What Is Life?

"I shall not today attempt further to define the kinds of material I understand to be embraced within that shorthand description; and perhaps I could never succeed in intelligibly doing so. But I know it when I see it."

Justice Potter Stewart, 1964

What is living? A definition of life might seem like the most fundamental bedrock on which a field as expansive as biology is built. You might therefore be alarmed to discover that there is no standard definition. At school, we get taught variations on a checklist for identifying the characteristics of living things:

- Movement
- Respiration
- Sensitivity
- Growth
- Reproduction
- Excretion
- Nutrition

Schools in the United Kingdom sometimes summarize this as the acronym: MRS. GREN. As a checklist, it works perfectly well for all the living things we see around us: you're probably doing at least five of those things right now.[1]

The criteria above are all facets of life, and are biochemical by nature—biology as enacted by chemistry. Movement, for example, can take many forms in living things. You scan these words when a network of specialized proteins contracts inside muscle fibers attached to your eyeballs, pulling and pushing the focus across the sentence. That's a very different type of movement to the daily bend of a sunflower as it tilts toward its solar power supply. That is controlled by a kind of inflatable joint at the base of the flower head, whose cells swell up with water on the side opposite the brightest light and bend the stalk toward the sun as it traverses the sky. And that's different again from the movement of certain self-propelled bacteria, who come fitted with a beautifully evolved rotor, the flagellum, which spins at up to a thousand revolutions per minute and can whip the cell along at a tenth of a millimeter per second.

We know very well that life is made of cells, and there are no life-forms that are not built from cells. Yet that is not what defines it, in the same way that a house is not defined by its bricks. We know that all life works via the reproduction of a universal code we can translate and read, but you won't be able to understand a baseball game by reading the rulebook. That point where the earth changed its status from "vitally inanimate" to "hosts living things" didn't happen by going down a checklist. There is certainly nothing wrong with that vital inventory; all of those examples on the checklist are

---

1. For an even more reduced stock of animal behavior, a more memorable but less specific list is sometimes referred to as the four Fs: Feeding, Fighting, Fleeing, and . . . reproduction.

by definition biochemical. The way proteins are built in strings and folded into three-dimensional thickets to give cells specific abilities; the way DNA can encode and replicate information; the way animal cells can take in oxygen, extract energy, and pump out carbon dioxide; the way plant cells can take in carbon dioxide, extract energy, and pump out oxygen. And so on. All of these processes are chemistry in action, determined by the behavior of the atoms that make up the molecules. So what is the nature of the chemistry that is actually biology?

Searching for a Definition

We have a checklist of the actions of life, but no clear, singular definition. Alone, or even in multitudes, a protein is not alive, nor is DNA, nor is a metabolic pathway. These are, for desperate want of a less clumsy term, inanimate. They do not have the breath of life in them, though they are clearly essential for it. And yet we are in no doubt that what allows an organism to live is the concatenation of all of these chemical events, and the ultimate termination of these chemical processes results in death. When we talk about the origin of life, everything pivots on the journey from inanimate matter to living matter.[2]

The transfer of information from cell to cell and from organism to organism via DNA is elemental to life as we know it. So is it possible to secure a definition of life that is underpinned by the ability

---

2. The word *inanimate* is unsatisfactory, as it implies inaction, which chemistry most certainly is not. The opposite of living—dead—is also unhelpful for the same reasons. *Undead* seems a fitting way to describe what is more correctly called prebiotic chemistry: fizzing, active reactions that form the pathway to life.

to transfer information? NASA's interest in life extends from looking for it in space to synthetically engineering it here at home, as
we shall see in part II. Astrobiology forms a satellite branch of
origin-of-life research, the search for traces of life in far-off places.
This new field is a coalescence of several branches of science—
geochemistry, biochemistry, astrophysics—to contemplate the
chances of life in the universe. NASA poses three questions in astrobiology: How did life begin and evolve? Does life exist elsewhere in
the universe? What is the future of life on Earth and beyond? The
first question—the origin of life—is of central importance to the
astrobiologist, as it will determine how we will look for and recognize life when we find it.[3] NASA's objectives in these outer-world
explorations have led them to adapt a definition of life in order to
help specify mission parameters: simply put, "if it's Darwinian, it's
alive." This is a position held by Gerald Joyce, a chemist working at
the Scripps Research Institute in San Diego, whose experiments are
precisely focused on allowing Darwinian evolution to occur in
DNA's cousin—RNA. He told me:

> If there's a sine qua non of life, it's the ability to un-
> dergo Darwinian evolution, and to have history in mol-
> ecules. Chemistry doesn't have history. Biology has
> history. To me, the dawn of life is the dawn of biological
> history written in the genetic molecules that are carved
> through Darwinian processes.

---

3. They've been doing this for a while. In the 1970s, NASA sent the Viking missions to Mars, two insect-shaped spaceships that sampled the Martian surface looking for the signatures of life, and found none. In August 2012, *Curiosity*, a rover the
size of a VW Beetle, gracefully plummeted to the surface, with one item on its ongoing mission being to scratch around in the red rocks looking for traces of what
once might have been life on Mars. It is the freakiest show.

This definition is focused in its entirety on information. DNA and RNA are codes, storing up a digital manual of instructions that can be copied and recopied over and over again. And as there is infidelity in copying, they acquire errors, which in turn become new information to be passed on and selected if it is useful. That is evolution.[4] Replication is crucial for life, but is it enough to count as a definition? A crystal can grow, and can replicate its structure. That growth can suffer imperfections but, according to this definition, it is not alive because those imperfections are not passed on. It is not Darwinian because it cannot acquire new characteristics via selection. Darwinian behavior is certainly an essential feature of all life that we see, but is only one part of a set of behaviors that are universal in life.

Jack Szostak picked up a Nobel Prize in 2009 for a career's worth of work in human genetics. He helped build new techniques for us to explore our own genome, and contributed to the discovery of key components of aging. Then he switched from the human branch at one of the tips of the tree of life to the almost utterly unrelated field of biology at the base of the tree of life. At Harvard, Szostak founded the Origin of Life Initiative. One of his many concerns is the spontaneous formation of the cell membrane, which we will explore further in this book. He is quiet, easygoing, soft-spoken, patient, and thoughtful. But when I interview him, it's clear that the quest for a definition of life exasperates him, and more important, gets in the way of researching it. Uncharacteristically, he *almost* snaps at me when I suggest that the absence of a comprehensive definition of life is problematic. He cuts me off mid-question: "I don't think it's a problem at all. I think it's completely irrelevant. What we want to understand is the pathway, how we went from really simple

---

4. Joyce's research into this area is utterly crucial, as he deals with the very origins of this process of descent with modification, and we'll come to it in the next chapter.

chemicals to more complicated chemicals: from really simple cells to more complicated cells to modern biology. We just want to understand the process and all the steps. We don't need to say, 'Here's the dividing line: on this side there's chemistry and on this there's biology.' The important thing is the path."

J. B. S. Haldane, who earlier in the twentieth century had helped construct the idea of prebiotic soup, took a similar line in 1949 with his straightforwardly titled book *What Is Life?* Chapter 14 bears the same title as the book itself but begins with this bold caveat

> I AM NOT GOING TO ANSWER THIS QUESTION. IN FACT, I DOUBT if it will ever be possible to give a full answer, because we know what it feels like to be alive, just as we know what redness or pain or effort are. So we cannot describe them in terms of anything else.

In other words, *I know it when I see it*.

Nevertheless, many people do focus on establishing a definition. Humans like to categorize things, science in particular, because so often it aids comprehension. Edward Trifonov, a biologist from the University of Haifa in Israel, recently approached the problem with an unusual tactic: to look at the words used in the many attempts by scientists to define life. By throwing them into a melting pot, he then ladled out a purified phrase: "Self-replication with variations." This is similar to the NASA definition and focuses on the transfer of information from one generation to another. I'm sure this approach was well-meaning, but it is hard to see the value in establishing a consensus definition based on the language used. Trifonov's publication was gracious enough to include many responses from a cohort of scientists disagreeing and, in Szostak's case, resisting the urge to constrain life into a definition at all.

Indeed, all such attempts run the risk of the "Blind Men and the Elephant" syndrome. The Buddha (but also deities in Islam, Jainism, Hinduism, and other cultures) recounts the story of several sightless men asked by a king to tell him what an elephant is like. Each examined different parts of the beast with their hands and concluded that a pachyderm was only the bit that they could feel. The tusk examiner suggested it was like a plowshare, the foot inspector a pillar, the tail man a brush, and so on. They fight; the king laughs. An elephant is all of those things, and you can't capture its majesty by singling out one of its characteristics.

When Justice Potter Stewart said "I know it when I see it" (quoted at the top of this chapter), the "it" he was referring to was pornography, in response to the prosecution of the 1958 film *The Lovers* to be banned as an obscenity. It has evolved into a phrase to describe the subjectively ill defined, or things without clear parameters. Life is just that. At one stage there was chemistry on Earth, and at a later stage there were living things. The route from the first to the second point is inevitably long, tortuous, and messy. The point where we definitely have living things is certainly where things become Darwinian, but not solely (as we shall soon see, there are even certain molecules that display the Darwinian property of self-replication with selection). In other words, the boundary between chemistry and biology is arbitrary. Life is a combination of lots of different chemical systems that are more than the sum of their parts. We separate science into categories in discussions such as this, and, indeed, when we learn it: biology, chemistry, geology, physics, et cetera. These are also somewhat arbitrary, as science is merely a way of knowing nature, and makes particularly blurry distinctions when considering the very inception of one of them: biology.

The Energetic Fly in the Soup

All of the actions of cells are ultimately mediated by the controlled flow of electrically charged atoms from one side of a membrane to the other. As you read, charge-bearing atoms flow into millions of single brain cells until they hit a threshold. When this happens, the brain cells fire up, and in consort with millions of others will form a thought process, or trigger a memory or understanding or induce the desire to make a cup of tea. Similarly, it is the controlled flow of protons—hydrogen atoms charged by being stripped of their single electron—across membranes that drives the generation of energy which the cell and organism entirely relies upon. In all complex life (including us), this happens in the cell's power stations, mitochondria; in bacteria and archaea, across a membrane just inside the cell's outermost casing. This type of chemical pathway is part of what we call metabolism, and is of central importance to all life as it generates energy. That energy powers all of the actions of biology, including the ones that facilitate the replication of information across generations, and everything else on the MRS. GREN's life list.

For this reason, in considering the foundations of chemistry as it relates to biology, we have to turn to a more fundamental science: physics. The way chemicals behave is determined by the laws that belong traditionally to this field. Ernest Rutherford,[5] the discoverer of the particles that make up the atom, famously declared, "Physics is the only real science. The rest are just stamp collecting." Though provocatively mischievous, there is some value in this reductionist view, one that is typically shared by physicists. Biological behavior

---

5. No relation.

is determined by chemical behavior, which is determined by atomic forces, and these are in the realm of physics.

Haldane's *What Is Life?* was published in 1949, but he was not the first to come up with this deceptively simple title. In 1944, Erwin Schrödinger wrote a physics-focused biological treatise with the same name, a classic text drawn from a series of public lectures.[6]

That this analysis should come from a physicist is perhaps fitting in further blurring the artificial boundaries between the modern disciplines in science. Physics, by its nature, tends toward the fundamental, and Schrödinger's conclusions derive from one of the absolute and most nonnegotiable universal rules: the second law of thermodynamics. This is the principle that dictates with total authority that over a period of time, energy will always flow from a higher state to a lower state and never in the other direction. We see the application of the second law all around us. Once the heat is turned off, a pan of bubbling soup can only cool down. This principle extends to every aspect of our lives: the heat of a radiator is dissipated into warming our rooms because it is attempting to equilibrate the two imbalanced energy states: one is hotter than the other. It will never happen in reverse. The measure of the second law is what we refer to as entropy. At a constant temperature, a plump balloon will only deflate unless its knot is perfectly sealed, in which case it will remain constantly inflated. But it will never expand if its surroundings remain unchanged. Within the closed seal of the balloon it has reached equilibrium and its entropy is constant. But

---

6. Schrödinger is best known for putting an imaginary cat in a box, and then poisoning it to death. Or maybe not. His famous thought experiment was a way of understanding the effect of observation on the quantum world and how it might affect the classical physical world. The cat, invisible to our observation, would be killed by the poison gas that has a random chance of being released. But we have no way of knowing whether or not the cat is dead until we open the box. According to quantum physics, until that point, the cat paradoxically occupies two simultaneous states: alive and dead, only to be selected upon observation. Perhaps less mind-bending was his attempt to focus on the first of those states.

relative to the rest of the world, it has a higher energy state and therefore wishes (if one can express desire in a balloon) to spread its energy more fairly. This tendency toward fair distribution in energy is represented by an increase in entropy.

What Schrödinger argued was that living systems are the continual maintenance of energy imbalance. In essence, life is the maintenance of disequilibrium, and energy as life uses it is derived from this inequity. It is sometimes described as a far-from-equilibrium process. The entropy of the universe is bound only to increase, thereby ultimately creating a more balanced but less ordered existence. The temperature throughout the universe will eventually be the same, following the even distribution of its total energy as the second law of thermodynamics commands.[7] Schrödinger recognized that all living organisms evade the decay to energy equilibrium during their life spans, and continue to do so in their descendants. This has occurred continuously on Earth for almost four billion years. We consume food, and energy is extracted from it inside cells. In doing so, we construct an order within our bodies, without which decay would transpire.

This order maintained in living cells appears at first glance to be in direct contravention with the second law of thermodynamics, which dictates that entropy will always increase, and therefore organization will decrease. Chaos is the ultimate direction of all things, and living things are not chaotic (at least on the terms laid out by physics). But this apparent paradox is not a problem. The law states that the incontrovertible increase in entropy occurs within a closed system. The universe in total is a closed system, as there is, by definition, nothing but the universe. On a more local scale, living things are not a closed system. We produce waste as a result of our

---

7. At this point, universal entropy will be at its maximum, and the universe will have reached so-called heat death. But you can relax for now, as this will not happen for many trillions of years.

life-sustaining metabolism, and expel it into the rest of the universe. While order is increased and maintained during any life span within the living thing itself, this seeming contradiction of the second law is more than compensated for by an overall increase in entropy beyond the confines of that organism: that is, your waste. The entropy of the amount of waste you have generated in your life is overwhelmingly more than the reduced entropy your body maintains by being ordered. And thus, the laws of the universe remain perfectly intact.

Life is a process that stops your molecules from simply decaying into more stable forms. The process of living is the chemistry that perpetually holds off decay. And this is precisely why the concept of the primordial soup is flawed. The notion that the right ingredients in the right surroundings might generate a self-sustaining life-form ignores the underlying principle that life is a far-from-equilibrium process. The chemical activity in a soup can only accede to the second law of thermodynamics: unless it has an external source to maintain an energetic imbalance, it will only decompose. In Stanley Miller's experiment, the spark of lightning may have triggered the formation of amino acids, but did not power a system of disequilibrium. Once those chemicals had reacted, they would do so no more.

Bill Martin (University of Düsseldorf) is one of the main critics of soup-based origin-of-life science, and we will come to his work shortly. He proposes an easy experiment to challenge the concept of primordial soup: mash up a life-form of your choosing to the point where any cellular resemblance has been destroyed, but all of the ingredients are still intact. This experiment effectively occurs every time a cell dies, but spontaneous resurrection from this soup, with all of the right ingredients, remains a myth. Any model of the very origin of life that does not take into account the necessity of a continuous flow and manipulation of energy is building upon something that is already dead. Stanley Miller's iconic experiment remains important, though its comment on the origin of life is limited.[8] It shows, elegantly and incontrovertibly, how biomolecules will arise from basic chemistry in the right conditions. However, it reinforces the view of life as merely an assembly of chemicals contrived into a thing that can reproduce. A primordial soup is not a vital and energetic mix, as it has no way of sustaining an imbalance of energy, whether it is in a warm pond, a pumice raft, a muddy volcano, or any of the other locations that have been proposed for the origin of life. A primordial soup is condemned as a midden: a decomposing dump.

In a sense, we living things are out of step with the rest of the universe. In the *Origin of Species*, Darwin described the "struggle for existence," meaning the fight to get food or a mate or the endurance against the elements. But it applies at a more basic level. To be alive is to struggle against entropy. Life does not violate the second law of thermodynamics, not at all. We cannot beat it, as that is the invincible force of scientific laws. In death we all submit to the will of physics, and our atoms accept their universal fate: to decompose and to be recycled, ultimately into less energetic states. Entropy strives to make the universe both more chaotic and more balanced. In doing so, entropy always increases; just like in casino gambling, the house always wins.

But by being alive, we have the opportunity to take something back from the house, or at least slow down its inevitable victory for a while. Life has evolved to extract energy from our surroundings and use it to maintain our vital information against the universal slide toward equilibrium by swapping and pumping protons from one side of a membrane to another inside a cell. Our lives, all lives, conspire to manipulate the fundamental forces of nature, and strive

---

8. The gasses of Miller's experiment were probably not the ones present in the early earth: it's now thought that carbon dioxide flooded the skies, and nitrogen was also present. That's a side issue here; we will never know the early earth's exact composition.

to do so continuously and forever. Jack Szostak is probably right, and tearing our hair out trying to encapsulate the essence of a life is merely a distraction from trying to trace back the path we know it took. But this indulgence into physics is crucial for understanding the origin of life. We know we can't go back in time and observe it. How it once happened is lost to science and history, so we do what we can to emulate it in conditions that gave rise to life's enduring imbalance. At some point, that imbalance acquired or created a system that allowed energy capture to be sustained independently. In that hatchery, the beginnings of encoded Darwinian descent could begin. But life-forms are first and foremost a sophisticated collection of chemical behaviors underwritten by the need for energy, and that informs how we go about experimentally quizzing the origin of life. In other words, we need to hunt for the signatures of metabolism.

It is clear that Luca was a living thing from which all subsequent life arose, but already carried with it many features that are essential components of the life that followed, including metabolism and genetics. In our cells there are two parts to metabolism: the first is the digestion of molecules to release energy, and the second is using that fuel to make life-sustaining molecules, including DNA and proteins. That poses something of a conundrum. From Luca onward, cells do the things on MRS. GREN's list, and we will come to the assembly of such a sophisticated system in this book's final chapter. Yet each item on the list is dependent on a chicken-and-egg problem. DNA encodes the proteins that enact cellular functions including metabolism, and these functions enact the decryption of the code itself. We've seen how the code works and how it was discovered. We've seen how it is the backbone of evolution and the template for the richness of all life. So how did DNA come to be?

CHAPTER 5

The Origin of the Code

*"For all inferences from experience suppose, as their foun-
dation, that the future will resemble the past."*
David Hume, *An Inquiry Concerning Human
Understanding* (1748)

ebrew has twenty-two letters. English has twenty-six, and
Sanskrit fifty-six. Chinese languages use pictograms,
which number in the thousands, depending on how you
count them.

Life only has four letters in its alphabet: *A, T, C,* and *G.* Throw in
a few extra glyphs and the flexibility of this alphabet goes up a
touch. Linguists refer to these tweaks as diacritics: examples include
the circumflex in French (*Â*) or the German umlaut (*Ö*). DNA's diacritics come in the shape of a small molecular appendage, methyl,
which is simply three hydrogen atoms bound to a carbon atom, and
gets tagged onto *A* or *C.* Just as in languages, these annotations are
an essential feature of genetics, modifying the meaning by labeling
sections of genome for particular characteristics. Most significant,
this earmarking of the code marks out areas of DNA to be silenced,

as if leaving the text where it lies, but scoring through entire blocks to say: "~~Do not read this sentence~~."

Even with this type of flagging, the most generous number of letters in living code is much lower even than the Hebrew alphabet. Evolution has given us a comprehensive description of how the wild spectrum of species has arisen from this simple code, but very little about how it came to be. When pondering the origin of life, however one tries to define it, at some early point the origin of our alphabet is a key question.

It's not just the alphabet that is bafflingly conservative. It combines into a similarly limited vocabulary. As explained in chapter 2, in genes those four bases are arranged into triplets, each one spelling out an amino acid. When strung together, these form proteins. But there are only twenty amino acids encoded in the genes of all life-forms. Furthermore, the way amino acids are encrypted in DNA is studded with redundancy. There are sixty-four ways of arranging four letters into groups of three. Sixty-one combinations are used in genetics to spell out just the twenty amino acids (and three indicating a "period" signal to mark the end of a protein). This means that many triplets encode the same amino acid. For example, *TTA* spells out an amino acid called leucine; but so also does *TTG, CTC,* and three other variations.

This redundancy allows changes to be acquired in our genes that don't make potentially harmful changes in the proteins they encode. If, during the process of a cell splitting in two, a random DNA error switched *TTA* to *TTG,* it would still spell out leucine, and the protein containing that amino acid would be unchanged by this unforced error. If it changed the final *A* to a *T,* then leucine would be replaced with phenylalanine, an amino acid with similar properties. Such an error would therefore potentially change the nature of the protein, but perhaps not too dramatically.

We see some similar redundancy in language. In England your

favourite colour might be blue. But six thousand miles west, blue might alternatively be your *favorite color.* Through random mutation, almost certainly a copying error at some forgotten point in history, American English has excised a letter that Brits believe is essential. It is patently not, as the pronunciation is the same, and it's not difficult to imagine a world where my British extra *u* slowly degrades out of use altogether.

That's not to say that all mutations are equally benign. Drop the *r* from *friend* and you have a companion with an opposite character. Given the amount of DNA replication that goes on, harmful mutations are relatively rare, though it may not feel like it if one is unfortunate to suffer from a genetic disease, a category that includes all forms of cancer. When these types of mutations do happen, a single letter change can have catastrophic consequences. A single base change in DNA that results in a switch from one amino acid to a very different one may well cause problems. Switch the *A* to a *T* in the β–globin gene and the amino acids switch from glutamic acid to valine, which has quite different chemical properties. The result is a misshapen globin protein, which in turn deforms the shape of red blood cells into bent and elongated curves instead of the uniform round lozenges that they are meant to be. The host of this single letter copying error will have the disease sickle cell anemia, a blood disease that frequently causes premature death. Such is the harsh but mercifully unlikely nature of genetic disease.

For the most part, though, these errors are of little consequence. They are normal and happen every time a cell divides. In the copying of the whole genome during replication, proofreading does occur. The copying proteins, called DNA polymerases, check that the new strand they are making matches up with the template, ensuring that wherever there is an *A*, a *T* is set down, and not anything else. But they're not perfect, and sometimes during cell division new changes make it through the proofreading. If that occurs in a sperm or an egg, it can be the first step in the genetic game of telephone that drives evolutionary change. You are different from your parents not just because whole genes get shuffled around during sperm and egg production, but because those cells bear new single changes in their DNA, which will be random, and therefore unique to you alone. Sometimes the pairing of the As and the Ts, the Cs and the Gs goes awry, and one of the two strands of DNA bulges out like a mismatched zipper. If uncorrected by the proofreading proteins, this single change can be passed on, and may slightly alter the behavior of the protein it codes.[1]

How is it that we have settled on four letters, arranged into triplets? With such a neat, self-contained system, unpicking it is tricky, and knowing where to break into the circle is perplexing. One approach, though, is to work out the minimum amount of DNA required to encode the twenty amino acids that life uses to make all its proteins. If there were only three letters of genetic code, then there would be twenty-seven possible combinations of triplets, still more than enough to encode the twenty amino acids we need. But in reducing the redundancy by more than half, the buffer against potentially harmful diseases is also reduced. The three letters in the triplets are not all equal. We see patterns in the ordering of the bases in triplets and the amino acids they encode. The first base relates to the process of the source of the amino acid. Amino acids float freely in the cell, waiting to be assembled into their codified proteins.[2] Some

---

1. The frequently quoted stat that genetically all humans are 99.9 percent similar is of relevance here. That means that if we compare two individuals' DNA, they differ by only one letter in a thousand. But if you factor in the three billion letters of code in a human's genome, three million individual letters are different (and don't include all manner of different, larger variations within our codes). That is a lot of variables to play with, and goes a long way to explain why we are all unique, even identical twins. Your precise genome sequence has never existed before and will never exist in another person. Humankind is in the 99.9 percent, but you are encrypted in the wealth of the remainder.

2. A more detailed description of this key process is described in part II, as scientists are beginning to subvert and redesign this process to create new, unnatural proteins.

are the products of the cell's metabolic cycles, but we cannot make nine of them, and so we have to eat them. By comparing triplets with the same first letters, we can deduce their origin as being either self-made or eaten. The second letter corresponds to a type, options including water loving (they dissolve easily) and water hating (they don't). The first two letters have a clear purpose in determining the product. The third, with all its wild-card flexibility, secures the deal, locking it into just one of the twenty. We can therefore speculate that the first form of DNA code might have been not a triplet but a twin, fixing the deciphering to essential sets of manufacture processes. The addition of a third allows both more combinations and more variation in the sequence, creating a conservative code that not only protects against the impact of catastrophic change, but gently encourages subtle modification. In short, our DNA encourages evolution.

But the underlying code is frozen, without change for probably four billion years. Alphabets are ice-cold but not frozen. They do change, although the change is imperceptibly slow. DNA, though, is indeed locked, at least in nature—though that is changing in the new world of invented biology, as you will see in part II. Francis Crick once thought that the alphabet of DNA came to be fixed as a result of a "frozen accident": a system that worked well enough and either outcompeted other versions or was the only one. Yet it is now clear that the inequity of the letters in each triplet is no accident, and became frozen with a delicate balance between fidelity and mutation, like a loving parent encouraging children to explore but at the same time protecting them from harm.

Ghost World

We can begin to see how DNA might have originated in a simplified form and settled into its current stable existence. But the paradox here is much more complicated than the old problem of the chicken and the egg.[3] Copying DNA is dependent on proteins, and proteins are encoded in DNA. DNA is the code and protein is the active product. But there are very solid reasons to think that at the beginning of life, the origin of genetics began not with DNA, but with its simpler cousin RNA.

Again, by looking at how modern life-forms work, we can infer earlier, cryptic roles for the mechanics in cells. RNA is the middle part of Francis Crick's misnamed central dogma: DNA makes RNA makes proteins. In order to understand how this sequence might have come about, it's useful to think of it in steps. We know of no way to directly make protein from DNA without the go-between RNA, so we could suppose that the first part, "DNA makes . . . ," might have been added after the concluding step "RNA makes protein"—RNA being the coded transcript from which the proteins are made. RNA, a single strand, is less chemically stable than DNA with its paired helices; it is more prone to falling apart. DNA retains the code for proteins with one strand reflecting the other, providing a mirrored backup service: where there is an *A* the other strand holds a *T,* and where one holds a *G* the other bears a *C.* Therefore, it's not unreasonable to imagine that DNA emerged later as a more secure form of data storage.

More recently, a pathway for that transition has even been suggested by looking at how error prone DNA and RNA are when copying one to another and back again. A Harvard team lead by Irene Chen compared how faithful the copying is when making RNA from RNA, RNA from DNA, and DNA from RNA. This is akin to testing the prowess of an Internet translation engine by inputting a

---

3. Obviously, egg-laying reptiles came hundreds of millions of years before the evolution of any birds, let alone *Gallus gallus domesticus*, the humble chicken.

sentence, translating it, and then back again to see how mangled it gets. Unequivocally, using DNA as a template came out on top. When copying RNA from a DNA template, the transcript was most faithful. This suggests that a move from an RNA world to the one we have today could have occurred without static—the message would be preserved. But DNA copied from RNA was riddled with errors.[4] With this concept in mind, it seems reasonable that as an information-storage device DNA is more robust, more secure, and more faithful than RNA. The transition from a world where RNA was the information bearer into the biological era that we know is referred to as genetic takeover; once that annexing had occurred, it seemed we could never go back.

That chicken-and-egg conundrum—DNA codes, protein acts—is at least partially solved with RNA as well, as on occasion RNA can do both. A second clue for thinking there was a world sporting the signs of life that was populated by RNA, not DNA, comes from the cell's protein factory, the ribosome. When it is time for a gene to make a protein, the process goes like this: the gene, set in DNA in the host's genome, receives an instruction to be turned on. A protein unscrews the double helix, like unwinding a wire cable tie, and separates the two strands. Another protein clamps onto the coding strand (the other is a mirror, and performs minimal function) at the three letters *ATG*. This codon is for the amino acid methionine, but also marks the beginning of a gene—the start codon. From there it walks mechanically forward, copying the DNA into a mirrored

---

4. For example, the sentence "But DNA copied from RNA was riddled with errors" is translated into Icelandic by Google Translate as "En DNA afrita frá RNA var riddled með villa," and back into English as "But DNA copy of RNA was riddled with errors," which is inelegant but pretty close.

Repeating this experiment with Czech results in "Ale DNA zkopírován z RNA byla prošpikovaná chybami" and then "But RNA copied from DNA was riddled with errors"—exactly the opposite of the original meaning.

RNA molecule: where it sees *ATG*, the RNA will read *UAC*.[5] When the transcription of the DNA is complete, the wispy RNA molecule floats off, bearing the message of the gene, and is helpfully called messenger RNA. The ribosome then picks up this transcript and ingests it, one letter at a time. It reads the code three letters at a time, each codon specifying an amino acid, which is delivered to the ribosome from the cell's milieu. As each codon is read in turn, the ribosome picks up the amino acids and strings them together to make a protein, which is expelled into the cell, dispatched to serve its purpose.

The ribosome itself is made of several parts, just as many of the active parts of life-forms are.[6] These smaller parts, by jiggling themselves into position, self-assemble far more easily than flat-pack furniture. But what is interesting for our purposes is that more than half of those parts are made not of protein, but of RNA. These long, heavily folded slivers of RNA link together with proteins to make the working ribosomes, and function just like proteins in that they enact a process. With these types of RNA molecules, therefore, we have both information and function. With RNA as the forefather of DNA on the early earth, the paradox of DNA and protein—the former encoding the latter and the latter making the former—vanishes. This idea is referred to as the RNA world hypothesis. We can get around the chicken-and-egg problem by not needing either. At some ancient juncture from mere chemistry to biology, the central dogma of "DNA makes RNA makes proteins" was simply "RNA makes."

---

5. RNA uses uracil as a replacement for thymidine, which differs only by the presence of a small cluster of atoms called a methyl group, also used as the diacritic tag in DNA methylation.

6. For example, the transporter of oxygen around our bodies, and what makes blood red, is called hemoglobin, which is assembled from four units of protein parts called globin, around an atom of iron.

The First Photocopiers

*NNNNN*NUGCUCGAUUGGUAACAGUUUGAAU
GGGUUGAAGUAU–GAGACCGN*NNNNN*

Can you see the family resemblance? This is R3C, the child of Gerald Joyce and Tracey Lincoln from the Scripps Research Institute in California. If evolutionary genetics is the process of tracking the slight changes in our genes back through time to reconstruct the Darwinian replicators of the past, R3C might be the end of the line.

With RNA being a decent contender for the first world of information and replication, this is a simple piece of RNA that does both those things. The letters are standard RNA (with *N* being a wild card, representing any of the four bases *A, C, G,* and *U*), and is made of two parts (shown separated by a dash). In a test tube, R3C contorts into a hairpin shape. Its function is to make a mirror version of itself by linking the two parts together. This propinquity for its reflection in turn makes a new copy of the original, and so on. This goes on ad infinitum, as long as the system is fed with the ingredients that will enable the continued chemical reactions which drive replication. Hundreds of millions of copying molecules can be made in a few hours.

It is a kind of protogene made of RNA, not DNA. Unlike our own genes, which have functions and instructions for proteins that build tissue and bone, or issue commands so that other genes will do so, the instruction that R3C carries is simply "copy me." DNA relies on other mechanics to copy itself, but R3C doesn't need any help: it's a photocopier that only makes other photocopiers. The fact that its only instruction is to copy itself doesn't make it very useful for a modern organism. But genetics had to start somewhere, and, hypothetically, this might be similar to what the first genes looked like—a copying machine. RNA that has function is called a ribozyme.[7] It's in these simple, double-action molecules that many scientists see a fundamental precept of life: reproduction of information. Joyce believes this is "where chemistry starts turning into biology. It's the first case, other than in biology, of molecular information having been immortalized."

In biological terms, the ribozyme is the smallest unit of information; in computer terms, a single bit. But in the test tube, as they promote their own replication, they enact a form of Darwinian selection. Joyce's experiment at first was marked with fidelity: the ribozymes reproduced themselves impeccably. But evolution requires flawed copying, and, as Joyce tells me, "perfection is boring." So Joyce introduced imperfection into the mix—sequence infidelity, the equivalent of spelling mistakes—in order that each copy was potentially fractionally different. Hence, Joyce introduced the foundation of Darwinian selection: variation. Out of a pool of these short RNA molecules, ones that reproduce themselves with greater success emerge to become the dominant form. These molecules have, through no guiding hand other than Joyce's providing the right conditions and feedstock, undergone a nonliving form of natural selection.

It's a clever approach to the origin of the genetic code. These experimental ribozymes are not natural but they are only partially designed. David Bartel and Jack Szostak created a technique in the 1990s that gave a great fillip to the idea of an ancient RNA world before DNA and proteins. Their technique allows functional ribozymes to at least partly create themselves. It's akin to the fabled monkeys-with-typewriters idea; that with a big enough number of primates bashing away on keyboards, one will eventually type a

---

7. This is a portmanteau of *ribo-* for RNA and *-zyme* from enzyme—proteins that catalyze biological reactions. It is annoyingly similar to the word *ribosome*, which itself has ribozymelike properties—that is, RNA with function.

Shakespearean sonnet. Bartel and Szostak put together a pool of literally trillions of strings of RNA, which were identical for a short stretch at one end, then random for the next two hundred letters. In the trillions of possible combinations they were looking to find one that, purely by chance, had the ability to add another similar RNA molecule onto itself. They then fished in that pool with a tagged piece of RNA, a bait to hook any one of those trillions of random molecules that could clasp another RNA molecule. In the simian-typing metaphor, it would be like searching trillions of simian text files with the phrase "darling buds of May." If that sounds easy, remember that these are entirely random sequences, and they are asking for meaning, or in this case function, to come from nothing. Bartel and Szostak found exactly that ability from their pool of RNA molecules at a rate of around one in twenty trillion.

The monkeys-with-typewriters example is purely random, and makes the point that, when working with very large numbers, patterns (or prose) will emerge.[8] Evolution is very much not random. The variation in genetic code may have arisen entirely by chance, but selection (whether natural or by the hand of a creator) is quite the opposite. Bartel and Szostak fully emulate biological evolution with their method by allowing variation to occur, but selecting the variants that work. Then, just as in nature, they repeat, but this time using only the ribozymes that already had demonstrated an RNA linking ability (as if, hoping not to stretch my monkey metaphor too far, giving them the first line of a sonnet as a starting point to their frenetic random typing). After ten rounds of fishing for those that could bind RNA together, the ribozymes that had made it through this harshest of talent contests were several million times better at

---

8.   In 2003, researchers at the University of Plymouth in England tested this idea in a very small-scale experiment. Six macaques were left with typewriters for a month, and produced five pages consisting mostly of the letter s. However, they also destroyed the typewriters by urinating and pushing feces into the keyboards.

RNA joining. There's a drop of medicine to take with that, which is that naturally occurring enzymes work around ten thousand times faster than the best of Bartel and Szostak's ribozymes, which were unlikely to improve much more. But recall that we are dealing with introducing function into a world where there was none, and this RNA world was transient, destined to be replaced by the more efficient and effective DNA and proteins.

In essence, these experiments are attempting to create something more fundamental than life. That ribozymes were the first genes, needing neither protein nor DNA, is not something we can actually know for sure, though it is uncontroversial now. Ribozymes do currently exist in nature, but if RNA self-replicators were the first genes, they've been extinct for more than three billion years, and we are experimenting to resuscitate an unrecorded dead language that is only known by its descendants.

In Cambridge, England, the Laboratory of Molecular Biology (LMB) has been called the Nobel Prize factory, as nineteen of those most sought-after medals were won in a dull gray glass building. The LMB recently relocated around the corner to a shiny new gray glass building, where Philipp Holliger runs a lab to explore the lost RNA world. His crew used a similar test-tube evolution to pick out another candidate for an aboriginal gene: they started with a smaller lucky dip of a paltry ten million variants and, by using some technical wizardry involving oil and magnetic beads, selected a ribozyme that produces more RNA. This one doesn't replicate itself, but they managed to cajole it into producing an entirely different ribozyme, a well-known one rather more dramatically called the hammerhead.[9] So far, Holliger's ribozyme can only make new RNAs of around one hundred bases in length, which is a little short for a ri-

---

9. The hammerhead ribozyme is found in a number of viruses and single-celled organisms. Its name is derived from the fact that when the RNA sequence folds into a 3D structure, it has a head that is exactly the same shape as a hammer.

bozyme and a long way from the average modern gene, but Holliger and his crew are working on extension.

Here we have the emergence of plausible models for the origin of information in chemistry and, more important, the origin of information copying. This is a hallmark of life. The amazing ability of Joyce's and Holliger's ribozymes is their spontaneous function. Joyce's copies itself, and Holliger's copies a lot else. The great goal is to shape the emergence of one that does both.

Creation by Deletion

But we can go back further. Ribozymes, as with all genetic code, have their four letters with which they harbor information: A, U (instead of T in DNA), C, and G. If we assume that Luca had this alphabet in place, and that is as far back as our historical records can go, we have no record of how this complete code came to be. This is unlike the evolution of language, where we have historical records of previous forms with different letters and meanings. The acquisition of the language of DNA and RNA is crucial to the understanding of the origin of life. To think that all four letters of code were acquired simultaneously seems more unlikely than if they were acquired sequentially, one at a time. We can test that idea by deleting them sequentially.

As in Scrabble, some letters are more valuable than others. *C*, cytosine, bound to a *G* is not as stable as a bit of code containing *A* and *T*, which will be the first to fall apart in heat or cold. On top of that, in RNA, the letters can pair up in a slightly different way from the neat and tidy rungs of the DNA ladder. This is rather sweetly called wobble pairing, and helps the RNA fold into the loops and folds that give ribozymes their self-copying ability. In the absence of *C*, both normal and wobble pairing can still occur.

Once again starting from a colossal pool of ribozymes with random variations in their sequence, Gerald Joyce and his colleague Jeff Rogers drove evolution by repeatedly fishing for ones containing no cytidine,[10] using bait with little affinity for that letter, but that could still connect two RNA molecules. Effectively, they were breeding out a particular trait from their stock, but instead of its being an undesirable characteristic in an animal or plant, the stock is a 140-letter ribozyme, and the trait is the letter C. Even in the absence of one-quarter of the available letters, they bred a ribozyme that could ably retain its joining function.

Once you've discarded one of the letters and showed that biological molecules can still work, what, then, is the next experiment? Joyce's approach, obvious but no less brilliant, was to evolve a ribozyme with only two letters of code. Necessarily, this involved evolving ribozymes with similar but different letters, in this case referred to as D and U, and the activity of the molecule they bred was much reduced from a full alphabet version. But nonetheless it still functions as an active biological tool. It may even be the case that this limited alphabet molecule would have been advantageous in hot Archaean waters, as the folded bonds containing C might have self-destructed at high temperatures.

These clever experiments don't show what actually happened in the origin of genetics, but they do show what might have happened. They show that evolution can proceed with an alphabet that is more restricted than the one life now uses. They feed into a framework that persuasively builds up an origin for coded replication; in other words, genetics. If one of the defining characteristics of life-forms is the ability to store and reproduce information, then one of the most fundamental questions of the origin of life is how such a complex

---

10. Cytidine is the name for cytosine attached to the ribose sugar, that is, the rung attached to the strut.

system could have begun. How it compares with what might have happened four billion years ago is difficult to say. We have the advantage of ingredients and design, at least experimental design, rather than the messy, low-tech, bucket chemistry of the early earth. But that first time had millions of years, whereas scientists have made these new replicators in a decade. The task is not to replicate what happened once; that will be impossible. But in all origin of life studies it's important to remember that we know the answer: life is the answer. The question is finding a believable route to get there, and in these experiments such a thing is beginning to emerge.

The Origin of the Alphabet

We can go back further still. When Gerald Joyce asks one of his brews to copy and mutate ribozymes, he has to supply the ingredients. The Archaean earth didn't have a sterile lab with sterile glassware and manufactured chemicals bought purified from industrial companies. Joyce's ribozymes provide a plausible mechanism for the first genes, and the basis for a language that has survived for billions of years. The letters of this language are the bases of RNA, so the next question is: where did they come from? These are not trivial molecules, certainly in terms of their importance in bearing the code of life, but also because they are complex. We describe them as complex molecules for a couple of reasons, the first one being that the letters of genetic code are manufactured in a difficult-to-learn biological pathway involving many different proteins. Our cells do this unthinkingly, as they have evolved highly advanced processes of metabolism that sustain their own existence by making their own parts. We look at these processes and carefully work them out with awe, and sometimes terror, when studying for exams. The metabolic pathways that create bases in our cells are evolved over innumerable iterations—the most stringent, patient, and ruthless design process known. Yet naturally, if these complex chemicals predate the seed of life's tree, there was no sophisticated metabolism with which they could be built.

With that in mind, the chemical structure of the letters of genetic code is complex primarily because we say so. If that sounds like circular logic, the second reason we think of them as complex is because they're not easy for us to synthesize chemically, either. The cell's process of synthesis is intricate so we see it as complicated, and our attempts to replicate that synthesis are tricky as well. Those bases have to line up in exactly the right way to form a working ribozyme, and for decades this has proved to be largely elusive in the chemistry lab. Additionally, there is a major supply problem. To make one of Gerald Joyce's self-replicating ribozymes you need seventy-odd RNA bases, but that's only if you know exactly what you are making. If one were to emerge from a random pool of millions, you'd need billions of bases to begin with. And when it starts copying itself, the number of letters goes up exponentially. Ten rounds of replication of a 100-base ribozyme would need a pool of more than 10,000 bases, but one hundred rounds of copying would need more than 10^{30}. Add to this the fact that each round reduces the concentration of the letter pool, which makes the replication harder. It would be like spelling words out of alphabet soup; after using up a few letters, it becomes harder to spell more words. In order for you to keep spelling out new words, you need a constant supply of letters, possibly an entire Campbell's factory's worth.

Therefore, before we get to the self-catalyzing, self-copying ribozymes, an earlier, more basic issue to examine is where those ingredients come from in the natural, inanimate, embryonic world. The problem is this: without biochemistry to hand, how did these letters spontaneously form, when we find it so difficult in the lab? This has been a question for nearly forty years, the duration of the RNA world hypothesis's life.

If DNA is a twisted ladder, the two essential components are the rungs and the struts. In RNA, which is only a single strand, it is like a ladder cut vertically in half. In order to get the individual letters of code to link up into working molecules, they have to link up to the struts of the ladder. This third component is a linker made of phosphate—a combination of phosphorus and oxygen atoms. The struts, meanwhile, are made of types of sugar molecules: the *R* in RNA stands for the sugar ribose, and the *D* in DNA for deoxyribose, which is the same except for one less oxygen atom. Each molecule of these sugars is attached to one of the letters, *A, C, G,* or *T* (or *U* in RNA); and when these sugar molecules are stacked up, they form the backbone, the ladder's struts. In DNA two backbones pair up, in RNA they remain as a single strand. It is the shape of these molecules that determines their behavior. For example, uracil is a small, hexagonal ring of atoms, and it links up with its sugar, ribose, which is a pentagonal molecule. Together, they make up a single rung of RNA, and these will get linked together by phosphates with the other bases to make a piece of genetic code. This brief aside into possibly painful chemical structures helps us to realize why recreating life anew is a tricky business. Only these precise shapes will act as part of living cells, but there are plenty of other ways these molecules could be rearranged. An atom in the wrong position, a molecular appendage in a different place and you don't get DNA or RNA; you don't have life.

The trouble is that biological chemistry in cells just happens, whereas we have to try quite hard to emulate it. While we can observe and deconstruct the mechanism that cells use to manufacture these molecules, for the purposes of studying the origin of life we need to understand their synthesis before those manufactories were present. Outside the foundries of our cells, getting the rings of sugar and uracil to link up has proven to be a pain.

Chemistry is an ancient science and has well-worn pathways of synthesis. Making complex molecules is frequently done by laying one stone at a time, building up to the final product. Some steps are harder than others, and atoms have to be cajoled into forming bonds where we want them to be rather than somewhere easier. Gentle coaxing can persuade two reluctant atoms to link up on the pathway. Yet in the case of making uracil for use in RNA, this stepwise chemical synthesis has proven unfruitful, as it has a reluctance to link its two component rings.

John Sutherland and his team are also at the LMB in Cambridge. His approach to this problem is a bold step, at least in terms of chemistry dogma. In a landmark study in 2009, Sutherland successfully bypassed the obstruction by taking an entirely different route.[11] The traditional approach is to keep the production of the two components—the ribose rings and the uracil ring—separate, as both processes produce a tar of products. Mixing them together, it was assumed, would result in a chemical sludge containing many sugars and a lot of other stuff, but not much uracil. To get around this, Sutherland ignored the prejudice of that stepwise pathway to sludge and decided to mix the ingredients together initially. This technique is called systems chemistry, in which instead of assembling the parts of a complex molecule one after another, and purifying them after each step, all of the ingredients are mixed at the same time. It is in some ways reminiscent of Stanley Miller's experiments in the 1950s, which produced amino acids in abundance from a stock of ingredients supposedly present in the young earth. It's categorically different, though, in that it is not merely mixing together ingredients to see what will emerge. The production of uracil in this way is heavily designed to do exactly that, in an environment that conceivably was present. It's not often that scientists use a word such as *plausible* in

---

11. This work was conducted while Sutherland was at the University of Manchester, another Nobel Prize factory, boasting more than twenty laureates.

the title of a paper, but in Sutherland's case it fits their model: "Synthesis of Activated Pyrimidine Ribonucleotides in Prebiotically Plausible Conditions." I have been dismissive of the idea of primordial soup in the previous chapter, on the grounds that it is not energetically plausible for a self-sustaining living system to emerge from a chemical swill. This is different, as it is a mechanism for building up the elements of living things, in this case the language of genetics.

Sutherland's approach works—uracil is formed—and is summed up with his comment, "Complexity is in the eye of the beholder." His reasoning is that it must have happened in the past without the assistance of a chemistry lab or the cellular machinations of biology. This new pathway is itself of great importance, but somewhat reserved for chemistry wonks. It turns out that the key missing seasoning was phosphate, which had the effect of jamming sugar production, but nurturing the linking of the two rings to form uracil.

Yet the approach says something significant about attitude in science. Sutherland's thinking is rather radical. For the first time, instead of thinking about the chemical conditions of the early earth and trying to emulate them to make uracil, he and his team figured out a way to synthesize uracil and then postulated that these conditions were likely to have been the chemical swill from which code built itself. After all, we already know the outcome on the early earth: it happened, regardless of how hard we find repeating it. Uracil and all the elements of genetic code were created somehow and somewhere, and if we can recreate that, we potentially know more than we did about what the early earth was actually like.[12]

---

12. There's a coda to Sutherland's synthesis. He and his team found that shining ultraviolet light onto the mix had the double effect of improving the yield of uracil and of carving up some of the by-products. UV light on Earth comes from the Sun, stronger now than it was when our solar system was young. Naturally, this is a plausible addition to the Archaean reaction, as these conditions were present on young Earth, but makes the idea more focused, at least in a location. If this actually transpired, then it happened on Earth's surface.

Our prejudices about the difficulty in nonbiological creation of complex molecules was also challenged in 2008 when a team led by Zita Martins at Imperial College London isolated uracil in the Murchison meteorite. That 220-pound rock had fallen to the Australian earth in 1969 and has been subject to scrutiny ever since, not least because it's big, which means there's lots of material to work on. It's also full of carbon, and so of interest to astrobiologists, many of whom suggest that extraterrestrial rocks might have delivered some if not many of the chemical components to Earth. Certainly, other bases and amino acids have been found in meteorites; given the flow of these boulders from space during the Late Heavy Bombardment, this might be a plausible supply. Either way, the presence of extraterrestrial uracil shows again that its synthesis is perfectly possible no matter how hard we find it to do.

It's difficult to shake people's prejudices, but there is no place for dogma in science. When explaining his new pathway, Sutherland faced vocal criticism from entrenched chemists decrying the simultaneous formation of the ingredients, insisting on the traditional view that synthesis should be incremental and stepwise, not derived from a chemical brew. He now jokes that this criticism is the "Dem Dry Bones" model, in which each element is created ready to be assembled into a whole: "The foot bone connected to the leg bone." So how does evolution produce a lower limb? By connecting a preformed foot and leg, of course: "Now, hear the word of the Lord!"

With these experiments the emergence of the universal language of life is not solved, but they reveal a pathway that looks rather plausible. The language of life is a staggering example of natural design. Not only does it have the ability to encode the wonders of the living world, it is built with a natural buffer in it—one that encourages evolution, but not too radically. We now have a grasp of how this code came to be. We have credible roots from basic chemicals to the letters of the language, we have simple RNA molecules that act like genes and promote their own replication, and we have a reason to think that DNA would be a better data storage device and so would ultimately replace the RNA world, where the genetic code could have plausibly originated. The conundrum at the origin of the central dogma—DNA makes RNA makes protein—is partially undone by the heroic ability of RNA to do the jobs of the others in that equation: code and function. With ribozymes, the mechanics of gene duplication is handled by RNA itself; it doesn't need the proteins that perform this vital function (as well as others). But the broader chicken-and-egg problem remains. Replication in cells from Luca onward requires energy. Therefore, underlying the RNA world, the transition to DNA, and the establishment of cells and life as we know it, must have been a source of energy that could be captured and manipulated to form the basis of all the energy-vampire processes that cells need. And here, we find ourselves in hot, deep water.

CHAPTER 6

Genesis

"If you come to think of it, what a queer thing Life is! So unlike anything else, don't you know, if you see what I mean."

P. G. Wodehouse, "Rallying Round Old George"

It's a heavy load, but you wear it so casually. You read these words with the accumulated weight of humankind's achievements behind you. For all the wonder of the genetic and cellular circuits that allow you to function, accumulated over billions of iterations in the unbroken lineage of cells, we're also a species that is the product of our culture. We invent, compose, trade, share, learn, and create. But more important, we carry these things with us through families, through groups, and across our species. Cultural evolution is not restricted to the descent of genes through pedigrees. We can learn from anyone else and acquire their skills without a genetic bond. There's a lucrative intellectual trade in declaring that one of the things humans do is what defines us as a species. It wasn't the invention of tools that switched us from simpler ancestors to modern humans, though that was part of the process. Earlier human species constructed simple tools, such as flint chippers or crude ax heads. Plenty of animals use tools, too, from otters' using stones to crack mussels to crows' wielding sticks to dig out a fat grub from a log. What's important is not that we do uniquely human things, but that we continue to do them. Humans, above all, *accumulate* culture.

You are the sum of your parts and our species, genes passed from parent to child for generations, and culture and ideas passed from human to human in all sorts of directions. This should be obvious, just as there wasn't one moment that made you who you are. We love to pretend that there are lightbulb moments in which something flipped and a character or an event was forged—Rosa Parks choosing to remain seated on a bus; an unnamed Chinese student defying a tank. They are iconic moments but don't describe the flow of history. Life—your life, any life—is not merely a series of incidents. It is the accumulation of everything you experience.

The beginning of life was just that, too. The transition from chemistry to biology was the accumulation of the things that life does—feeding, copying, reproducing, and so on. At some point in the earth's history there were just chemicals, and at a later point there was life. The most exciting question in science is how that transition happened.[1]

There are many components to the rebuilding of life from the bottom up, and the quest for a simple answer to a problem of immense complexity will probably be futile. How it happened the first time and how it will happen the second, and subsequent, times will be different. But the challenge of origin-of-life initiatives is to

---

1. Even death is not easy to define, for the same reason that life isn't. We can point to events that are essential for death, those that are in opposition to life—loss of respiration, the stopped heart, snuffing out neural activity, the end of consciousness. But none of them is useful to the majority of life-forms that don't have brains or hearts or consciousness. Doctors use a checklist to pronounce death, but death is a process itself, not a punctuation mark at the end of a life.

emulate creation in plausible and credible scenarios. That means thinking not just about what the processes were that transformed chemistry into biology, but also where.

From Foreign Shores

Why should we have such a conservative system of biology: one code, one mechanism, one central dogma? We now know that RNA probably acted as a precursor to the more robust system in place today. Francis Crick speculated that the universality of DNA was a frozen accident, a system that worked and was locked into place, forever it seems, to the exclusion of any alternatives. Once in place, any changes would either be lethal to the bearers of a novel vital system or outcompeted rapidly by existing life. Why did the code become fixed? The truth is that we don't really know. Studies have worked up ideas that it was no accident, and that the four-letter code and the twenty-amino-acid lexicon are optimally balanced at a point where positive and deleterious mutation permits evolution to occur successfully. There is a seemingly simpler explanation, one that Crick was briefly seduced by. It is this: the code didn't evolve at all. It was delivered here, already functional but frozen, from somewhere else. In this case, "somewhere else" means outer space.

This is one of the vaguely more scientific versions of an origin-of-life idea called panspermia, which maintains that the earth was seeded with living things at some point in the deep past, and evolution of species ensued. Cast aside the images of space aliens, belligerent gray bipeds with spindly bodies and bulbous heads. The aliens described by panspermia's advocates would be

simple, bacterial-type cells, or even simpler, the mechanics of cellular life to be harnessed by whatever the Archaean planet had available. Carried on comets or meteors, they would act as the seeds of
evolution, an infection from the stars. Upon successful delivery,
they were free to engage in evolution by natural selection.

Francis Crick, writing with Leslie Orgel, authored a paper entitled
"Directed Panspermia" in 1973, not just outlining the possibility of
extraterrestrial delivery of the seeds of life but also arguing that they
were sent here deliberately by intelligent beings. Crick and Orgel were
drawn to the idea of directed panspermia largely because of the inadequacies of existing ideas about the origin of the genetic code. It's a
strangely tongue-in-cheek study, contemplating aloud the limitations
of space flight and speculating with wise foresight on the discoveries
of planets beyond our solar system. At that time none was known; at
the time of this writing we are up to over a thousand.[2]

It's not difficult to see the appeal of panspermia, though scientifically, it is neither plausible nor credible. Even today, in this era of
local galactic calm, around thirty thousand tons of debris descends
onto our planet each year, mostly in the form of dust disintegrated
in the atmosphere. Showers can be seen burning up in the sky at the
right time of year, such as the Perseids in each northern hemisphere
summer. We even know of interplanetary relocation, at least in tiny
quantities.[3]

Once in a blue moon, a big one hits, like in Murchison, Australia,
in 1969. Mercifully rarely, a colossal one lands, like in Chicxulub

---

2. Crick and Orgel's paper is written partially in the formal language of scientific
papers, but also drifts off to wonder if the fate of the alien bioengineers was that
their star had "frizzled them." This is not a term that appears frequently in academic journals. It's unusual to see such august scientists pontificate in public, in
what looks to be the formalization of a fascinating bar conversation: "The
psychology of extraterrestrial societies," they note with presumably arched eye-
brows, "is no better understood than terrestrial psychology."
3. Very occasionally, a piece of Mars lands on Earth, most recently in 2011, a
33-pound meteorite named Tissint after the small town in Morocco it landed in.

sixty-five million years ago. These rocks are from outer space, formed
when larger space rocks collide and send their debris on a crash trajectory with Earth. These events are now uncommon, especially in
the case of the carbon-rich meteor that landed in Murchison, loaded
with molecules that we know are essential for life on Earth. But leading up to our best estimates of Luca's Earth was the intense and relentless meteorite activity of the Late Heavy Bombardment. Millions
of tons of space rocks were deposited here over a period of millions of
years, to the extent that a small proportion of the earth's mantle is
made up of interstellar rocks. It would potentially only take the survival of one single cell to seed all that followed.

Alas, the attractiveness of an idea is of no consequence in science. It's the data that counts, and therefore it's easy to deal with
panspermia as an explanation for the origin of life on Earth. There
is simply no evidence for it. While it is appealing in a sci-fi sense, we
have no evidence for life ever having survived beyond the orbit of
the moon, and there it was only us as tourists.

That is not to say life does not exist beyond this pale-blue dot
we call home,[4] but we have never seen it. Not in the claims of
alien abductees, nor in the microscopic bobbles of rocks from
Mars,[5] nor in the credible scientific hunt by SETI, the Search for
Extraterrestrial Intelligence. The rate at which we discover Earth-

---

4.   As Carl Sagan described it following a photograph taken by *Voyager 1* in 1990,
3,728,227,139 miles from Earth. Many now think life elsewhere in the universe is
statistically inevitable. Indeed, the blossoming field of astrobiology exists to for-
malize the study of if, how, and where life might exist extraterrestrially.

5.   A paper in *Science* in 1996 suggested that the shapes of particular microscopic
mineral deposits on the Allan Hills meteorite (commonly abbreviated ALH 84001)
were biological in origin. The rock in question is thought to be around four billion
years old, blasted off the surface of Mars in an impact. Some researchers maintain
that the bobbles seen on the meteorite's surface are the product of microscopic life-
forms, but the scientific consensus is that they are geological in origin. Chemical
analysis of the rock shows that the amino acids it bears are identical in nature to the
Antarctic ice it was found in, suggesting this was the very terrestrial source of biological molecules. Research continues.

like planets is rocketing as we get better at looking for them. But for the time being, we are alone.

Certainly, we see many of the ingredients of life in space. We are, it is important to remember, in space, formed in space, and in formation, part of the solar system. To ignore this is to deny the fact that our existence is determined by the space in which the earth exists. Often, the press reports a new discovery concerning the extraterrestrial origin of one of Earth's constituent parts. Invariably, these studies are reported, with excitable glee, to be extraordinary. The true revelation of these discoveries is that they were previously unknown, not that delivery via comets or meteors might be inherently surprising. We now know that many of the ingredients of life, the component parts, are present in space, including amino acids and even, as discussed in the previous chapter, elements of genetic code. These are important scientific discoveries, especially for the study of the origin of life. They show that the chemistry which produces biological molecules occurs outside biology. That means that synthesis of those components is not limited to Earth. And it describes a mechanism of delivery, whereby comets or meteorites can bear these chemical parts and potentially deliver them to our surface.

Earth has continually and unsurprisingly (though inconsistently) been the recipient of things that fill our local space. If we make a clear delineation between "Earth" and "not Earth," then the ingredients that make up our living planet are going to be manufactured in one of these two places. Is it therefore a surprise that water, methane, and other chemicals that figure as elemental standards in biochemistry originated off-world?

Panspermia is seductive because it looks like a simpler explanation, one that does not need the patchy, incomplete time line we currently have in place. But in fact it's not simpler; it's simply a dodge. If there were evidence for the transfer of life from elsewhere in the universe to Earth, and there were no other plausible explanations, then panspermia might have some legs. But it doesn't address the question of how life started on Earth, and simultaneously provides no evidential support for its starting anywhere else. Parsimony enforces the notion that it might look like a simpler explanation than our very incomplete alternatives. But it is far from being the case, because it requires not only that life abounds or at least exists in the universe, but that it is also based on DNA. We simply have no evidence to support that. As it is, there are other, better avenues worth exploring; they have their own problems, but at least they are scientific routes, ones that can be tested and retested. Reliance on extraterrestrial life as the delivery boy for a cosmic egg is neither necessary nor sufficient for the origin of life. For now, and for the foreseeable future, life from beyond our pale-blue dot must remain in the dominion of science fiction.

Containment

All cellular processes, including the mode of information replication, are underwritten by a system that harnesses and uses energy, and we will come to that shortly. But first, there is another issue to deal with. Cells are discrete units of life, either in an organism or free-living, and so their information and metabolism have to be kept separate from the rest of the universe. We therefore have the problem of the origin of the membrane. Containment of the guts of a cell is an absolute part of being alive, and it is easy to think that the cell membrane is merely a container, a sheer balloon in which the machinations of life are contained. Certainly, containment of a cell's vital organs is essential, as it has the effect of concentrating chemical reactions that maintain life. But cells exist as part of an environment, not apart from it. The cell membrane is more akin to a custom house, an interface that monitors and controls the import and export of all the goods and messages a cell needs. This immense complexity allows an individual cell to exist in its environment or with the other cells in its host creature. This exchange comes in the form of a highly complex network through which traffic constantly flows, even when resting, and is carefully regulated through gated pores, pumps, and channels that stud the surface of a cell like seeds on a strawberry.

Cell membranes are built from fatty molecules called phospholipids. These look like split pins used for holding sheaves of hole-punched paper together: they have a water-attracting head and two legs that repel water. At Harvard, Jack Szostak's work is centered on how the cellular membrane came to be. He experiments with simpler molecules with only one tail, to see how they behave, and how they might help us understand the formation of the first cells through self-organizing behavior.

When Szostak mixes together a delicate blend of these fatty molecules, they do something that is simultaneously remarkable and not. They self-organize into a tiny bubble, about the same size as a bacterial cell, a hundredth of a millimeter wide. It's remarkable to us, because they look a bit like cells spontaneously forming out of straightforward chemicals. It's unremarkable, for the reason that they do this is because they can.[6]

The fatty acids have a bulbous head and zigzaggy tail, and that gives them their split personality. The atoms in the head are

---

6. Spontaneous self-organization is not quite as magical as it might seem. If you had Cheerios as your breakfast cereal this morning, you will have seen fundamental universal forces conspiring to invoke spontaneous organization in your bowl. The oat rings want to float, because their density is lower than that of the milk. Gravity pushes them down, but the pressure underneath from the column of milk pushes them up. If, after the first few mouthfuls, there is enough space at the surface, they will automatically jiggle themselves into a hexagonal pattern, because this is the formation that allows the upward force to be distributed evenly.

arranged so that they like butting up against water. The tails are quite the opposite, and repel water. But the tails also attract one another. In the right solution, a watery one, they therefore jostle together into two regimented lines, tails facing inward, and heads in the water. At the right concentration, this line will expand in all directions and link up to form a sphere. The nature of those fatty acids is such that they are chemically content when organized into this kind of membrane. As soon as that happens, there is an inside and everything else.

Modern cell membranes are studded with pumps, channels, aerials, and receivers to ensure a healthy contact with the extracellular world. Biological mailboxes lie embedded in the membrane to receive input signals from around the body and local environment, and strong anchors link up neighboring cells to hold tissue together. All of these membrane tools maintain living order within the cell and within the organism. Szostak's simple membranes are a long way from this highly evolved interface. But the first cells would not have access to that complex machinery. Part of the challenge of engineering spontaneous self-organization is to start with simple molecules and evolve more complex behaviors upward. Szostak calls his creations protocells, and many of the things they do are similar to what cells do. If you feed them, they grow and divide. They spontaneously absorb simple genetic molecules, short fragments of things similar to DNA. When they do, this absorption triggers growth and fission. If you heat them up and cool them down, they not only survive intact but DNA inside also undergoes a type of replication, which can stimulate growth and division of the whole.

It's thought that these bobble-headed phospholipid molecules would not have been difficult to find on the early earth, not because we have traces of them, but because they're easy to make with different recipes. One of the most striking is in experiments that blast icy fragments resembling the composition of comets with ultraviolet light to mimic an interstellar forge. But they've also been found in meteorites, and made with much more Earth-bound reactions. As far as ingredients go, there is ample sufficiency, and they make a good contender for the first membrane. But as the origin of life was chemistry transitioning to biology, these prototype membranes need to acquire complexity from their simple, self-organizing origin.

The membranes of modern cells need to be sophisticated to prevent leakiness. A free flow of immigration into a cell can cause all sorts of turmoil and unrest; similarly, accidentally spilling your most prized secrets and possessions is a dangerous game. So the custom weigh stations on the modern membrane preside over a careful coordination that will expel molecules (for messaging or as waste) and let the right ones in. It is not enough to say that the complex system we see now is simply better than what came before. In order to uncover the real transition, you have to show why one is preferential to the other. In evolution, things are never frozen and rarely accidents. So why would a complex system take over from a simpler one? Itay Budin and Jack Szostak experimentally asked that question of their protocells. The answer seems to come from that most Darwinian of concepts: competition.

For natural selection to play out its hand, variation is needed. If everyone was the same, not only would it be quite boring, there would be no potential foothold for advantage. Getting one up on your siblings is what that variety produces if you are lucky enough to bear a competitive advantage. It could be for sunlight, food, or a mate, but without competition there is no change. Even though Darwin didn't deal with the origin of life other than idly speculating on it. Not for the first time, we see a Darwinian process playing out in experiments that are most certainly prelife. The variation added to the stable mix of simple protocells in Budin and Szostak's experiment was the addition of minute quantities of complex phospholipids more like the ones in modern cells. They nestle into the membrane and sit quite comfortably next to their single-tailed colleagues. However, those protocells grow one-sixth larger than their simple brethren. A cell that acquired the ability to make these phospholipids would therefore immediately grow larger than one that could not. The increased size also encourages cell division.

Szostak's experiments, by his own admission, are a long way from replicating the spontaneous formation of anything as sophisticated as the contemporary cell membrane. There are all sorts of caveats to assuming that this is the re-creation of what happened the first time around. The phospholipids need to be made somehow, as they are not abundant on Earth or in the heavens in the way that their single-tailed cousins are. That requires at least the most basic form of metabolism, which speculatively, in an early biochemistry, might be supplied by mechanisms hitherto undiscovered. On top of that, the winning cells, with their modern phospholipid membranes, lose their ability to let molecules inside and out as they please. That demands the emergence of the channels and pores which make the modern cell membrane a custom house, rather than merely a port. Protocells are apparitions of cells and do cell-like things. The simplicity of their formation is great solace for those who find the spontaneous emergence of all the facets of cellular life a big ask. Containment of living processes is necessary but not sufficient to enable life as we know it, though these experiments suggest a pathway. Nevertheless, the spontaneous formation of protocells in a dish is a simple yet significant step in re-creating genesis.

Protocells also mimic cell division. In us, the process of a cell splitting in two is a highly orchestrated and active maneuver that requires metabolic energy. Not only does the DNA in its entirety have to be copied, it has to be arranged for the physical split, and the membrane has to grow big enough to make two cells from one.

Modern cells have a network of internal struts and scaffolding that helps everything into its right place so the split happens evenly. At the beginning of cellular life, there was presumably no such architecture. Szostak's work also demonstrates that physical properties force cell division, as simple as squeezing the protocells through the pore in a filter. So again, as with the experiments that show how ribozymes might have once been the original code, the formation of the cell membrane is a lot less mysterious, thanks to these experiments that show how something sophisticated might have emerged spontaneously from something simpler.

Deep, Hot Water

In living cells, DNA and membranes—the information and packaging that sustain life's great descent—all require energy. The emergence of membranes and code can happen in a variety of locations if the ingredients and conditions are right, but if we accept that underlying these chemistries, life is a process of harnessing energy from the environment, a very different real-world scenario is required.

On a gray bench, on the first floor of a building in which Charles Darwin once lived, is a glass jar. It's about big enough to fit a human head inside. It is perched on a wide-legged tripod, and a tinfoil-insulated rubber hose is piped into its base, where near-boiling water rushes in. On one side of the jar, cold water flows in through two glass nozzles. The top of the jar is a heavy glass lid held in place with bolts and a thick rubber seal.

This is a bioreactor, which may sound fancy, but in fact is rather old-fashioned, like something drawn by the British cartoonist and illustrator William Heath Robinson. The laboratory itself, in the aptly named Darwin Building on Gower Street in central London, is filled with University College London's phalanx of evolutionary biologists and is a typical contemporary working hive of molecular biology with benchtop spinners, shakers, machines, and countless tiny tubes of colorless liquids. The bioreactor jar sits atop its tripod sprouting wires, tubes, and foil, looking somewhat ectopic in its otherwise hi-tech, sleek surroundings.

The magic touch is what is in the jar. About the size of a truckle of uncut Stilton is a block of what looks like gray stone. In fact, it's a ceramic, carefully designed to mirror the mineral growths that poke out of a newly discovered breed of submarine hydrothermal vent. It looks like a cross between a sponge and pumice. In fact, UCL's Nick Lane, who designed this rig, experimented with both of these when trying to generate the right amount of flow-through around the material. In baking this block of sophisticated clay foam, the bubbles and pores were delicately controlled to create just the right conditions to emulate the underwater bubbling vents that Lane believes are the best candidate for the site of the first life.

The word *vent* suggests vertical chimneys, as seen in the breath-taking images of the teetering towers of the Lost City way down at the bottom of the Atlantis Massif, halfway between the Canary Islands and Bermuda. Pictures and the description of this underwater metropolis were first published at the beginning of the twenty-first century, a new type of seabed hydrothermal field, fizzing with the energy of reactions between the mantle rocks and seawater. And despite heat of up to 194 degrees Fahrenheit and highly alkaline waters, these towers are teeming with life, so much so that lead scientist Deborah Kelley from the University of Washington in Seattle told *Nature* in 2001, "You can't even see the rock because of the amount of bacteria."

It's these vents, ever changing through the continuous effervescence from the active earth below, that Nick Lane's ceramic is modeling. The vents emerge from the seafloor as the plates of the earth shift and split, revealing fresh hot virgin rock drawn up from the mantle. Once exposed, they react with seawater, splitting it apart into oxygen and hydrogen and a host of other gasses that are equally brimming with energetic potential, eager to react. In so doing, these gasses percolate through the rocks, driving a honeycomb into them as they cool in the surrounding seawater, a process called serpenti-nization. It's in these tiny hatcheries that Nick Lane, Bill Martin, Mike Russell (at NASA's Jet Propulsion Lab at Caltech in Califor-nia), and a small handful of other scientists think the first living processes occurred, before RNA, DNA, and cell membranes. In the lab, the flow of water and gasses around the porous ceramic is not simply from bottom to top, but continuously circulating in and out of tiny pores, just as it does in the Lost City.

A modern cell's energy currency comes in the form of a molecule called ATP. It is a molecule that is constantly being made and recy-cled, as it holds energy in its chemical bonds that cells can use, to the extent that each day you will make and use your own body weight in ATP. The process by which this happens is a complex metabolic cycle that relies on there being a gradient of electrically charged hydrogen atoms (protons) across a membrane, a maintained imbalance. In our cells, that power generation happens in the mitochondria (the pow-erhouses of complex life) and in bacteria and archae, in membranes just inside cell walls. Special proteins on these membranes act like turbines, and the flow of protons through these turbines results in the energy being generated, which gets stored in ATP, which is used to power all the processes of the cell. This cycle is so fundamental to so many living processes that it seems like a good contender for the most elemental metabolism life can have, and therefore a system that could underlie all life. It relies simply upon there being more protons on one side of a membrane than the other.

This is why Mike Russell has driven the idea that the hydrother-mal vents of the Lost City offer up a model for the nursery of first life. Due to the precise qualities of the chemicals bubbling up from the reactions between the rocks and the sea, they form natural proton gradients in the swirls around the honeycomb rocks. In our cells, that proton gradient is maintained by our biochemistry, and it is the maintenance of that imbalance that keeps us from drifting toward decay.

The heat of the vents is not what is important: we don't use heat to generate our energy, nor, indeed, a bolt of lightning as Stanley Miller's experiment did, so why would first life? Cells harness chemical energy in many forms, with almost inscrutable metabolic pathways, but with ATP produced by the flow of a proton gradient at its core. These are systems that are far from equilibrium, and just like life, a continuous restriction of an unfettered increase in entropy.

In a sense, the bubbling vents are a mixture of the right ingredients in such a way that biology can emerge from chemistry, but it is a far cry from the spontaneous generation of a primordial soup. In the vents it's the circulating gasses flowing out that maintain the thermodynamic imbalance. Specifically, protons stream around the pockets in the rocks, to be concentrated into the alkaline interior away from the acid sea. At the bottom of the Atlantic Ocean is a bubbling bioreactor with a thermodynamic imbalance of streaming protons preventing retirement to equilibrium. The metal nuggets studded in the rocks catalyze the whole process, and the empty cells of the porous rock concentrate the ingredients in this never-resting mix.

As a site for the origin of life, hot vents fit seductively well, and Nick Lane's bioreactor is one of very few experiments to test that idea. It is ongoing, and at the time of this writing results are unknown. If successful, Lane hopes to see the emergence of reactions and chemicals that are similar to or the same as the ones we see in biology. This might then suggest a place for Luca's birth.

But Bill Martin pushes the model even further. To him, Luca is

126   CREATION

something that is not a free-living cell at all.[7] Even with a bushy network between single cells that populated the earth for two billion years, Luca still stands as a common ancestor of both archaea and bacteria. In Martin's version, Luca was a swill of activity, but not a free-living cell as we might have assumed. The beginnings of life were not inside a membrane-bound cell at all. Instead, the last common ancestor of the two oldest domains of life was locked inside the rocky shell of the alkaline underwater vents. In there, concentrated and protected, but fed and with a constant supply of charged atoms, is the transition from chemistry to biochemistry, and then on to life. The evolutionary split that wrenches archaea from bacteria happens after the RNA world has given way to DNA, but before the arrangement of fatty molecules into the cell membrane. They have the same genetic code, and they both rely on ribosomes as their protein factories. In both, they are more similar to each other than to our cells. But in archaea, the tools for writing and copying RNA—a protein called RNA polymerase—are more similar to our own than they are to bacteria's. The archaea's outermost shell, the cell wall, is different from a typical bacterial cell wall, and the membrane is radically different from anything else. These differences are present in contemporary species, but they are fundamental enough to have roots beneath the bottom of the intertwined bushy bit of the tree of life. According to Martin, first life was housed not in membranes, but in rock.

Martin's model starts with bubbling hydrogen, ammonia, and

---

7. This is a heterodox view from a man who likes a healthy bit of agitprop. He starts his lectures on the origin of life by tipping up the soup bowl. He points out that Haldane and Oparin, the independent founders of the soup model, were both Communists. Coming from a Texan (who now lives in Germany), this instantly sounds like a political point. In fact, he is quick to point out it means that, as they were both dialectic materialists, "there had to be a rational explanation for everything from the workings of society to the needs of individuals to the origin of life. And [Haldane and] Oparin's theory became party doctrine."

hydrogen sulfide frothing around the pores in a serpentinized rock. These pockets look rather like the empty cork chambers that Robert Hooke saw with his microscope at the very beginning of this journey—not cells as we know them, but cells like the honeycomb cavities of a beehive. The acid sea and the alkali interior of the vent rocks provide a natural proton gradient for the flow of power that incepts a basic form of metabolism, and it is continuous for as long as the vent is active. Simple biochemical reactions that we observe in modern cells begin to occur in this energetic mix, and we start to see amino acids being forged in this tumult. Next come other foundation biomolecules, such as sugars, purine, and pyramidine rings, and other molecules that we see in metabolic cycles. Purine and pyramidine go on to fuse, becoming bases—the letters of genetic code, maybe as they do in John Sutherland's syntheses. When these letters link, the RNA world can begin, only to be eventually replaced by DNA with its superior data-storage prowess. But it's all still trapped in the tortuous labyrinths of the rock, the pores concentrating biochemical action, rather than the dilution of a warm pond of a broth. This locked reactor is the last common ancestor of everything, rock-bound for now, but soon acquiring the skin that will allow it to break free.

It's here that life's first great schism occurs. Two sets of molecules grow to line the inside of separate rocky chambers that hold this brimming biochemistry. One will lead to bacteria; the second, to every other living thing. The biochemistry that has developed in these cells is shared, but from now on, as they evolve new powers, they build different types of membranes to house themselves, and new ways of maintaining the energetic difference between outside and in. There, stuck to the side of a gassy rock at the bottom of the ancient sea, cellular life begins.

This is a model. It is Bill Martin's plausible description of how genesis could have occurred based on what we know about simple life and the chemistry and geology of the vents. It's impossible to put a timescale on this transition. Roughly, we have a window of several hundred million years, somewhere between the earth-pummeled Late Heavy Bombardment and the fossil cells we can actually see, around 3.6 billion years old. Modern alkali vents are fairly stable, but rare. They may have abounded on the seafloors of the capricious young earth. Chemical reactions tend not to dawdle. Allowing a million years for two things to react gives ample opportunity for them not to—to be washed away. But this timescale provides the opportunity for the experiment to take place billions of times, over and over again in billions of pores, with infinite variables being tested over and over again. It is a conspiracy of chance made possible by the sheer numbers on offer. It only needs to have worked once, the jackpot in the chemical lottery.

We can't repeat the timescale, nor can we experiment in vents themselves. But we can model them. This is why Nick Lane's humble bioreactor is such an important experiment. It is an experiment in the purest sense, too, as he really doesn't know what will come out of it. The pores are seeded with iron sulfide (aka fool's gold), a catalyst that is present in the vents, which may speed up the energy transduction that is the beginning of metabolism. Just like in the vents, protons bubble up from the base and circulate through the labyrinthine pores. Each of many different biological processes is being tested individually, and the products examined. Lane is looking for the hallmarks of biological molecules—the bases of RNA and their component parts, amino acids that make up proteins, and the molecules that come from energy-harnessing metabolism. It is designed to find hints of biology built without enzymes, and at the time of this writing, this simulation is just beginning. The key to this experiment is that it is all far from equilibrium, just like life is,

and just like the vents are. It won't make cells, but it might spontaneously generate signatures of the chemistry that underlies all living
processes.

Although the definition might be elusive, we certainly know
what life is when we see it. Life is physics, chemistry, and biology.
That makes the attempts to recreate its origins a renaissance science.
That also makes it contentious, because different experts will approach from the angle that makes the most sense to them. The quest
to build life anew is to understand all of the things that life does, and
to re-create them individually at first, then gradually align them and
fuse them. Many of the pieces are being forged, all with different but
vital characteristics: energy, information, reproduction, metabolism, evolution.

The journey from simple chemistry to more complex chemistry
to biochemistry, from simple information to profoundly inscrutably
sophisticated genetics, and from simple soapy bubbles to dynamic
custom houses of membranes might seem unlikely. The astronomer
and author Fred Hoyle once described the unlikelihood of the ani-
mation of chemistry as follows: "The chance that higher life forms
might have emerged in this way is comparable to the chance that a
tornado sweeping through a junkyard might assemble a Boeing 747
from the materials therein." He goes on, in typically bombastic
terms, to say that it is "nonsense of a high order." In his book *Evolution from Space*, he made a calculation about the probability of the
proteins of a most basic living cell arising spontaneously, and came
up with the number one with forty thousand zeroes after it.[8] Hoyle
believed in panspermia and used this calculation to dispute the

---

8. Hoyle was a brilliant scientist and author but also something of an iconoclast,
prone to vocally rejecting mainstream ideas. He disputed the universe's origin's
being the result of the big bang, which is the overwhelming scientific consensus
view.

origin of life's being the earthbound transition from chemistry to biology. It was a fallacious argument and has been refuted sensibly many times. Hoyle's errors include the assumption that modern proteins were the first in life. He assumes that the trial of building a working cell was sequential, not simultaneous. He might have been contemplating the twenty amino acids that emerged from the sludge of Stanley Miller's famous brew, all present and correct, ready to fall into line into a working, living entity.

But in the works of Gerald Joyce, Jack Szostak, John Sutherland, Nick Lane, Mike Russell, Bill Martin, and others, we now see emergent properties that are not improbable at all, and are a lot less mysterious than we once thought. The conditions of the infant earth, tested in modern labs, render self-creation unavoidable.

We see RNA whose origin is random sequence fulfilling the roles of early enzymes and early genes. We see cell-like bags summoned into being by nothing other than atomic forces. And we experiment to see the hallmarks of metabolism emerge from the effervescence of inanimate chemistry. As we march toward an increasingly sturdy model of genesis, the mysteries of creation once filled by deities inevitably give way to science. In the lab, at least, we see a foreshadowing of the process that has driven the spread of life for four billion years. Ribozymes undergo a form of Darwinian selection, as does the transition from simple membranes to complex. This is not to say that Darwin had miraculous prescient knowledge of chemistry a hundred years on from his death. It's merely that the process of natural selection that he first described is more powerful than he could have possibly imagined. And "whilst this planet has gone cycling on according to the fixed law of gravity," as he wrote, a transition happened, driven by the power of a process that would one day lead directly to you. As we continue to explore the world of the cell, we are on the brink of seeing the start of this grand journey of life

on Earth, the only living planet we know of. And such humble beginnings, too. Every muscle twitch, every breath, every thought,
emotion, and sensation you've ever had, right down to the wince of
pain from an insignificant paper cut, started its journey in a microscopic chamber at the bottom of the sea, four million millennia ago.

PART II

The Future of Life

CHAPTER 7

Life, Not as We Know It

"What I cannot create, I do not understand."

Richard Feynman, 1988

What is the best way to find out how a thing works? There are a number of approaches, including observing how it behaves, or testing it (possibly to destruction or at least till it fails). But there is only so much you can learn from these techniques. To understand a thing of complexity you have to take it apart. A car mechanic cannot understand the complexities of the internal combustion engine merely by thrashing it on a racetrack. To understand an engine you must take it apart and see how the parts fit together, what each one does separately, and how they relate to one another.

Since before the discovery of the most basic unit of life in the late seventeenth century, the cell,[1] biologists have spent plenty of their

---

1. Some scientists consider viruses, or at least some types of virus, to be living. They are problematic, as they have many characteristics of the more definably living cells, but lack the cellular machinery to reproduce themselves. Thus, they are parasitic on living cells in order to reproduce. The question "what is life?" is explored in chapter 4, but for simplicity's sake I have chosen to adhere to the consensus (which is not undisputed) that viruses are not alive.

time observing and testing how living things work. But a much more instructive endeavor has been to take them apart. Leonardo da Vinci was an accomplished anatomist who drew many beautifully intricate diagrams of the insides of animals and people in order to determine structures and functions within our bodies. In the early seventeenth century William Harvey deduced the mechanisms of blood circulation by dissecting the fine construction of veins, arteries, and the animal and human heart.

Fifty years later, Antonie van Leeuwenhoek had discovered cells, the things from which all life-forms are built. As microscopes improved, the guts of these tiny bags of living matter were systematically unveiled and, in time, removed from the protective membranes that encase their innards. In the nineteenth century mere observation was supplanted by chemical detective work, and the composition of the components of cells—proteins and other essential ingredients of life—began to be described.

It was during the nineteenth century, with these chemical techniques, that DNA was first isolated, though it was almost a hundred years before its importance and its iconic double-helix structure were discovered. In 1869, a young doctor named Friedrich Miescher was working in Tübingen, Germany. As a result of the ongoing Franco-Prussian War, he had ready access to the pus-soaked bandages of wounded soldiers slowly rotting in a local hospital. Pus contains a surfeit of white blood cells called leucocytes, their purpose being to fight the determined invasion of disease-causing interlopers at an open wound. In extracting various ingredients from the fetid bandages, he isolated a chemical that contained significant quantities of the small molecular group phosphate. This particular cluster of five atoms is not normally detected in proteins, whose chemical composition had been well studied by this time. Therefore, Miescher figured his extract was different. He called it nuclein, as it came primarily from the nucleus of the soldiers' leucocytes, a separate compartment in the center of the cells of all complex life. Protein names tend to end in the letters -*in*, such as hemoglobin and insulin, so it is tempting to speculate that Miescher was still thinking of it in similar terms. His work was never really followed up, and it wasn't until many years later, when nuclein was shown to belong to an entirely different class of molecule, that it was identified as DNA.

In the twentieth century, biology matured into the study of ever-smaller components of cells. DNA was characterized famously by Watson and Crick using the data of Rosalind Franklin and Maurice Wilkins in 1953, and Crick, along with dozens of others, spent the next few years decrypting how vital information was stored inside the double helix. The workings of our complete set of DNA, the genome, were the focus of the grandest endeavor in biology: the Human Genome Project. That venture finished in the first decade of the twenty-first century with a complete read-through of the three billion letters of DNA in an average human, though the mysteries of human genomics continue to be explored.

Since Crick and Watson's landmark, the second half of the twentieth century was the era of molecular biology, and by the end of it the majority of all scientific research on Earth broadly concerned the molecules of life—DNA, its equally important cousin RNA, and proteins. Understanding the molecules and the language of life and their universal nature has profoundly transformed all aspects of our understanding of living organisms. As we explored in part one, DNA and molecular biology were the final bricks that cemented our understanding of the nature of evolution and provided a unifying mechanism by which we can trace our way back into the deep past, and to a single origin of life, a stem from which all life has grown.

It has also spawned a new way to examine living things. As we will discover in the following chapters, our solid understanding of

the molecules of life has led to an era in which we can alter, manipulate, and effectively remix the basic genetic code of any living
thing. We collect these related fields under the umbrella term *genetic engineering*, which has already begun to transform aspects of
our lives. It has revolutionized the discovery of how diseases transpire, as we can alter the DNA of animals and cells to mimic those
diseases and use those remixed cells as a test ground for bold new
medical treatments.

If the biological twentieth century was concerned with taking
cells apart to understand how they work, this newfound understanding has also given us the ability to put them back together
again, but intelligently designed by us with specific purposes. Cells
are minute manufactories, evolved over billions of years to have
highly specialized functions—to construct bone, store memories,
convert light into electricity in our eyes, or swallow invaders that
cause us harm. They either are collected into a community performing in harmony to form an organism or, as is the case with most
cells on Earth (bacteria and their lesser-known cousins archaea),
live free as single entities.

The quotation at the beginning of this introduction, "What I
cannot create, I do not understand," is attributed to the great bongo-
playing physicist and all-around master of science and science communication Richard Feynman. It was his last message to his
students, written on his blackboard at Caltech at the time of his
death in 1988. It is in the reassembly of the language of life that we
are at a place beyond mere comprehension of living systems, one
where life-forms are engineered tools. Genetic tinkering became
wholesale engineering and now has evolved into a new field whose
purpose is solely to create life-forms that serve as tools for humankind. It is known by the contranym synthetic biology.

Definitions in science are often imprecise and frequently unhelpful. Synthetic biology means different things to different people, and

aspects of those definitions are discussed in the following pages. Scientifically, it is a direct descendant of genetic engineering, and often the overlap between the two is unclear. For that reason, I have looked at both in the course of exploring our newfound ability to create life-forms using parts of genes and cells taken from the tool-box that evolution has provided. And finally, as a postscript, I will explore the creation of entirely new molecules that fit into DNA but are not part of evolution's lexicon of new languages and new uses for the genetic code.

All of the creations of genetic engineering and synthetic biology are new to the inventory of life. Most of them feature minor modifications or small additions that add a new function to (or, indeed, knock an existing function out of) an organism. Almost all occur in bacteria, whose genetics are understood well enough that they have become essential equipment of the modern era of biology. Even so, synthetic biology is a new field, and many people may be unfamiliar with it.

In May 2010, the news was briefly—and unprecedentedly—besieged by a single synthetic cell. It was given a name, and photos of Synthia, as it became known, adorned front pages and television channels all over the world. After fifteen years of toil and roughly $40 million, the geneticist J. Craig Venter published a paper that described a bacterial cell his team had created. Its genome had been assembled not inside a parent cell, as with every other cell in history, but in a computer. The press swooned. One magazine put Venter as the fourteenth-most influential person on Earth, sandwiched between David Cameron, the current British prime minister, and the American politician Sarah Palin. Although it was a major landmark, Synthia was not quite part of the synthetic biology revolution that is beginning.

Craig Venter himself was keen to mark this event in grand terms. To illustrate our command over DNA and to distinguish Synthia (aka *Mycoplasma mycoides JCVI-syn1.0*) from naturally occurring bacteria, Venter and his team hid what video game players call an Easter egg in its genome, in the form of secret encrypted messages in the DNA of his creation. One of these was "What I cannot build, I cannot understand," which fractionally misquotes Feynman. An unnamed member of Craig Venter's team had obtained it from the Internet, a source that on occasion has been known to be wrong. Errors aside, this Easter egg quotation served the purpose of stamping an indelible "watermark" into this bacteria, proving that it was unnatural in origin. It also showed that DNA can act as a data storage device, a means of encoding information that is not biologically relevant. We shall come to this emerging field in the afterword. Overall, this whole story was an impressive demonstration of just how far we have come in our technological ability to manipulate and build DNA.[2]

Synthia owes its existence to the Minimal Genome Project. *Mycoplasma genitalium* is a minor parasite that causes minor burning and itching in infected men and women when they urinate. That's not what is interesting about it, though. Venter's interest in this cell is because it has a mere 517 genes, and a genome of only 582,000 letters of genetic code—bases (compared with 4.6 million bases for the more common lab microbe *E. coli*, or 3 billion for humans). After sequencing the first complete genome in 1995, Venter and his colleagues' second quarry was *M. genitalium*. This choice was no act of laziness, but the critical first step in Venter's next big idea. The

---

2. Synthia contained three hidden quotations, about which British satirist Charlie Brooker commented, "The geneticists spliced a James Joyce quotation into the DNA sequence. The unsuspecting genome now has the phrase 'To live, to err, to fall, to triumph, to recreate life out of life' written through it like letters in a stick of rock. In other words, it's the world's most pretentious bacterium." The quote was taken from the novel *A Portrait of the Artist as a Young Man*, but Venter and his team neglected to seek permission from James Joyce's estate, which is notorious for aggressively protecting their copyright. Executors of Joyce's estate immediately issued Venter a cease and desist letter. In fact, the copyright had expired, and no further action was taken.

plan was (and is) to determine the smallest number of genes and the minimum amount of genetic code required to sustain a living cell. Life originated with only a handful of basic genes, and they were copied and mutated for all subsequent life to flourish. But those first cells must have started with an entry-level set. Starting with the smallest known genome (at the time) made a lot of sense. The long-term plan was to establish the minimal amount of DNA required for a cell to exist and reproduce, so that they could use that genome as a foundation on which they could build new functions. Those functions were typically grand: for example, to address the environmental damage that our use of fossil fuels has brought upon the planet by building microbes that might produce hydrogen for fuel. There are a couple of reasons for building a cell from scratch rather than modifying an existing one. For example, cells can be controlled more precisely in defined conditions such as labs, and they can have singular purpose built into their genetic program so that the normal functions of cells are curtailed. But the primary reason is control: in principle, we can potentially control a synthetic cell far better than natural cells that we don't fully understand.

Synthia represents a proof of principle, that it is possible to construct an entire genome synthetically, albeit a very small one. And it is possible to get it into a cell, so that the cell functions and, crucially, reproduces like any other bacteria. More broadly, this saga shows the ambition of scientists to turn biology into an applied science, into engineering. It indicates that the molecular manufacturing required in synthetic biology is not at all easy, which is not surprising, given how young this field is. The protagonists of the purest form of synthetic biology have engineering as their key guiding principle and, even more specifically, the commoditized version of electrical engineering. These aims are not merely to investigate and understand how living processes function, but to re-create, re-mix, and build living organisms that address global problems.

With the publication of Venter's creation, the media brandished a curious mixture of teeth gnashing and misplaced triumphalism. The *Daily Mail*, a UK newspaper not known for nuanced reporting on scientific advances, ran the story with a headline that bellowed: "Scientist accused of playing God after creating artificial life by making designer microbe from scratch—but could it wipe out humanity?" The very straightforward answer to that question is no, as is inevitably the case when newspaper science headlines end with a question mark. Here is why: Synthia's genome was designed with fail-safe devices built into it. In general, these work in two ways. First, you can include genes that ensure your modified bacteria is only able to grow in a very specific blend of food that only the lab knows how to produce. Second, the genome from which Synthia's was copied and modified is based on a minor goat pathogen, one that causes udder infections. Venter's team inserted a chunk of DNA that renders the cell incapable of causing infection. Therefore, while it might be able to just about survive outside the lab in which it was created, it would not be able to thrive in its preferred habitat unless someone with molecular biology expertise further modified its genetic limitations. Could it wipe out humanity? No, but if a skilled geneticist was persistent and mean-spirited enough, it might well bother a goat.

The brouhaha was not limited to tabloid newspapers. Eminent professors weren't shy in joining in the hubris. Julian Savulescu, professor of ethics at Oxford University, told the *Guardian*:

> Venter is creaking open the most profound door in humanity's history, potentially peeking into its destiny. He is not merely copying life artificially . . . or modifying it radically by genetic engineering. He is going toward the role of a god: creating artificial life that could never have existed naturally.

Of course, a Friesian cow or Savoy cabbage are both life-forms that are extremely unlikely to have existed naturally. They have been bred over innumerable generations for the express purpose of amplifying desirable characteristics: milky udders or a crunchy taste. I doubt anyone would accuse your average farmer of creaking open profound doors.

Did Venter create life? In a very particular sense, yes. But in most others, no. Certainly, Synthia did not exist as a living thing until Venter's team booted it up in a plastic tube. Its genome was created on a computer and then in a machine fed with four bottles containing the raw letters of genetic code. It was not derived from the biological replication of an existing genome copying itself using the method that has served life for billions of years. But the sequence itself, modified with genetic fail-safe suicide switches and pretentious (mis)quotations, was replicated from an existing species. And the chassis into which the synthetic genome was placed was from an existing cell from a naturally existing species, one distinctly embedded in the firmly rooted tree of life. What Craig Venter and his team did was to re-create a life-form synthetically. That is undoubtedly a huge achievement in itself. It's another incremental step on the pathway to having total control over DNA, and our ability to manipulate life.

The second thing that Venter's endeavor reveals is how poorly understood and misrepresented this nascent field is. We have been tinkering with living things for tens of thousands of years, though never before with such molecular precision. The pace of progress has been breathtaking. A decade ago I finished my PhD in genetics, during which time I was doing minor acts of genetic tinkering in order to work out how the eye develops, and why in some families it goes wrong, leading to children born blind. The types of experiments I was doing then can now be done by undergraduate scientists and even hobbyists known as biohackers—amateurs, sometimes of school age, who mix the genes and DNA of organisms in garages on weekends. This in one sense is natural progress, where new technologies become normalized and the exclusivity that comes with expertise or invention is lost as they become standardized tools. But the pace of that transfer is daunting, and poorly understood outside the world of science. And in the right (or worse, wrong) atmosphere, people fear what they don't understand. As the story of Craig Venter and Synthia shows, extraordinary science can result in visceral reactions.

What follows is a snapshot of this embryonic field of the engineering of living things. The synthetic biologists have application in mind, engineered and constructed solutions to the problems that this planet faces, and, more incredibly, the tools to explore beyond our earthly shackles. Many of the applications of the creations described here are speculative, but they are nevertheless built with purpose in mind. With the advent of these creations, we stand on the precipice of a new industrial revolution. The wood, iron, and coal that drove the industrial revolution of the eighteenth and nineteenth centuries was drawn from nature—cut, mined, and extracted from the world. But in use they were rendered inanimate and dead. This time around, the agents of change are not just taken from the natural world—they are alive.

CHAPTER 8

Created, Not Begotten

*"I am against nature. I don't dig nature at all.
I think nature is very unnatural."*

Bob Dylan

F reckles looks like a perfectly normal kid. She has bright eyes, a healthy white pelt, and gambols happily with Pudding, Sweetie, and her five other siblings, exactly as you might imagine a young goat would. Until I fend her off, she's very keen on chewing on my trousers. To the casual observer, and to professional goatherds, she shows no signs that she is not a perfectly normal farmyard goat.

In fact, Freckles is utterly extraordinary. While almost all of her genome is what you would expect a goat to have, there is a foreign corner in her DNA taken from *Nephila clavipes*, the golden orb-weaver spider. The language of the DNA is exactly the same in both goats and spiders, as it is in all living things, but the translation of that particular piece of code is very alien to your average goat. It has been inserted at a very specific point in Freckles's genome, adjacent to a coded instruction that prompts the production of milk in her udders. As a result of this genetic intrusion, when Freckles lactates, her milk is replete with spider silk.

Freckles is the creation of a team of scientists from Utah State University led by Randy Lewis. I visited their lab, which is in fact a farm at the base of a daunting mountain range in Logan, Utah. Though a long way from the sterile, clean rooms of a typical molecular biology lab, this setting is apt enough; Lewis and his colleagues are in a sense the most advanced farmers in the history of our species. It is dragline silk they are interested in, the gossamer fiber that spiders use to swing from when descending. This material has stunning physical properties, with a combination of strength and elasticity unsurpassed by anything that humans can yet manufacture. The orb-weaver genus has existed since the Jurassic period, which began around two hundred million years ago, and during the epochs before and since evolution has conspired to bestow a product on these spiders that is so far beyond humankind's greatest efforts, but which the spider produces in bulk without a thought. We can make fibers with strength or with great elasticity, but not with both. That is why Freckles exists, the unnatural creation of scientists.

For more than ten thousand years, since the dawn of agriculture, we have identified appealing characteristics of the natural world and attempted to enhance and exploit those properties. This is farming, a process that is at heart the precise opposite of natural selection, though it harnesses the exact same means. Instead of survival determining what unintelligently designed adaptations evolve, we choose characteristics that we find attractive and breed species to optimize production of that trait, whether it is fruit or land used to grow crops; livestock for meat, dairy, or leather; or dogs and plants bred for behavior or aesthetics. Farming is evolution by design.

Until the era of genetic engineering, this artificial selection has been inherently limited by natural boundaries. Although notoriously difficult to pin down precisely, one definition of a species is a group of organisms that, when crossed with one another, can produce fertile offspring. It's not a wholly infallible definition, and the process by which that barrier is introduced—speciation—is one of the great questions in biology. But from the point of view of a farmer, the limitations of the species barrier are unsurpassable. In simple terms, you can't mate a pig with a cow.

Desirable traits are introduced over time by careful selective mating over generations, a fact well known to Charles Darwin. The opening chapter of *On the Origin of Species* is devoted not to his central thesis of evolution by natural selection but to artificial selection in pigeons. In this pigeon section Darwin is demonstrating that species are not immutable, that they can change over generational time. The birds he describes had been reared over thousands of generations by competitive fancy breeders to feature ever more preposterous prize-winning plumes, gullets, and feet, such that they had names like Trumpeters, Fantails, Tumblers, and Pouters. Through meticulous observation of their skeletons, and in direct opposition to the forcefully held opinions of the pigeon men, Darwin correctly revealed that these pigeons were all still varieties of the same species, the rock dove *Columba livia*.

With the discovery of the genetic code and with the advent of our ability to manipulate it, we are now capable of fully circumventing the limitations of evolution and radically violating the species barrier. Freckles, though by no means extraordinary in terms of genetic modification techniques available now, is a dramatic example of this biological infringement. Pigs and cows are both mammals, both even-toed ungulates and closely related in evolutionary terms with each other and with camels, giraffes, and hippopotamuses. Their common ancestors existed a few tens of million years ago, an evolutionary blink of the eye. Yet that distance is enough to forefend viable offspring if they were to attempt mating. This barrier is laid down by diverging genetic incompatibilities, though the physical congress is conceivable. A few closely related beasts do produce viable children, such as zeedonks, ligers, and hinnies, though these hybrids tend to be infertile and therefore evolutionary cul-de-sacs as they cannot themselves reproduce.[1]

The last common ancestor of a spider and a goat would have existed something like seven hundred million years ago, at a time when the beings that would acquire hard shells, such as insects and crustaceans, were evolving away from creatures with fleshy exteriors, such as the fishy or reptilian beasts that would eventually lead to us. As the tree of complex life only grows via diverging branches, spiders and goats have not exchanged genes or DNA since that time. Unlike a zebra and a donkey (or even a pig and a cow for that matter), sexual union of a spider and a goat is obviously physically impossible. But the DNA code they both carry is the same across all known life-forms. It is the same language, formatted in the same way so that the tools and mechanics of the translation of that code are blind to its cryptic meaning. Thus, if a spider dragline silk gene is introduced into a goat genome in a way careful enough not to disrupt essential biology, the goat's cellular machinery will produce spider silk without reference to its aberrant origin.

And that is exactly what happens in Freckles's udders. In spiders, dragline silk is made of short strands of protein, which align and self-assemble as they are pushed or drawn out of the spider's spinneret. Proteins themselves are long strings of molecules called amino acids, and the very specific combinations of amino acids in silk fibers mean that they line up and overlap one another, locking

---

1. Zeedonks (aka zebonkeys, zebrinnies, zebrulas, and zebadonks) are a cross between a zebra and a donkey. Similarly, a liger is a lion and tiger cross, and a hinny a cross between a male horse and a female donkey—as opposed to a mule, which is the opposite. Sterility in these unusual hybrids is in accordance with a principle called Haldane's rule, (after J. B. S. Haldane, who came up with the term *primordial soup*), which states that when a cross is successful between two species, one sex will be absent in the offspring, and that will be the one with two different sex chromosomes, that is, XY (males) instead of XX (females), in mammals. In birds and butterflies, this is reversed.

into continuous threads that have built-in give. Because of this, they have length, strength, and elasticity. Randy Lewis's goats produce the short silk proteins in abundance, which float freely in the milk. The fat is removed from the milk, which is then pushed through a high-pressure processor. From there, with just a glass rod, I lift a single thread of spider silk from the goat's milk, and the noses and tails of the shorter fibers link together as I draw them from the liquid. It is strong enough to wind onto a spool by the foot. As yet, this unnatural silk does not quite have the same elastic or tensile strength as that created by spiders themselves. The research continues.

Although seemingly bizarre, this process is effectively next-generation farming. Its physical properties make spider silk an extremely desirable substance for humankind. It is tougher than the man-made fiber Kevlar used in body armor and puncture-proof tires. It is also a substance that does not generate a significant immune system response and remains insoluble in water, two facts that the classical Greeks exploited when they used cobwebs to patch bleeding wounds. These properties mean that spider silk has great potential for ligament repair, for which medicine currently uses either slivers of flesh extracted from the patient's own muscles or tissue from dead bodies. Neither of these solutions is permanent, so we could do with a replacement ligament made of a biologically neutral substance that has physical strength equal to or better than our own ligaments.

It might seem absurd to say it, but spiders cannot be farmed. When kept in groups they have a tendency toward cannibalism. With our ability to violate natural species boundaries using the tools of modern genetics, we are able to harvest a product by placing its code in a farmable animal. Although Freckles the spider-goat is a striking example of our prowess with gene mixing, she is not at all unique. This is genetic engineering: the creation of unnatural life using evolution's bountiful toolbox.

The language of DNA is carried by all living things because life on Earth ultimately evolved from a single origin, a cell that existed around four billion years ago.[2] The language, code, and mechanics of all living cells are all the same. DNA bears a hidden language, which is decoded via a messenger molecule called RNA, which is translated into proteins. Proteins carry out all the functions of life, as enzymes that speed up essential biochemical reactions inside our cells to provide us with vital energy, or as missives such as hormones that carry instructions from one part of the body to another. Proteins form structures that hold our cells together, and they fabricate other structures such as bone, teeth, and hair. In short, all life is built from or by proteins.

Because of this common origin and universal process, the cellular mechanics of reading and translating the code of DNA are both blind and indifferent to its meaning, which means that the code of living things can be used to transgress the natural barriers of reproduction. Chimeras such as Freckles are called transgenic organisms, as the genes have been moved across the species barrier.

Scientific progress always relies on the development of new technologies. Humans excel at tool use, and we have always looked to nature for useful instruments. We see what is available to us and appropriate and modify it for different uses, as we have since we used bones as clubs. The natural world has always served as our toolbox. We chipped flint into crude arrowheads, smelted metals from rocks, and wore animal skins sewn with bone needles and plant twine. The technology that enabled the violation of the species barrier which genetic modification requires is also borrowed from nature's staggering bank of resources. They come from our most distant evolutionary cousins: bacteria.

In 1968, Hamilton Smith (then working in the Microbiology

---

2. This cell is known as Luca, the Last Universal Common Ancestor, and is discussed at length in part I.

Department at Johns Hopkins University in Baltimore) isolated the first of a set of proteins that launched the era of genetic engineering, and for this he shared the 1978 Nobel Prize for Physiology or Medicine. These proteins are known as restriction enzymes and, put very simply, they act like DNA scissors. In nature, a bacteria's restriction enzymes cut their own double helix as a means of protection against the invasion of viruses. As a virus carries only its genetic code, and none of the machinery to replicate it, its purpose is to hijack the host cell to reproduce. The infected cell translates the virus's code and unwittingly makes new viruses, releasing them en masse until the cell bursts. Restriction enzymes are part of the arsenal to prevent that hijack by editing out the interloping DNA.

The ability to cut DNA is of profound use to geneticists, as restriction enzymes do not snip the code randomly. Instead, they only cut when they recognize a precise row of letters in the genetic code. There is a huge range of restriction enzymes, with some cutting at common sequences and others at very rare pieces of DNA. Imagine that working in this book: a restriction enzyme that makes a cut when it sees the word *cell* would shred this book into fragments, as if blasted with a shotgun. A restriction enzyme that cut when it recognized the word *jumentous* would chop the whole text neatly into just three pieces, as that uncommon word appears only twice.

This naturally occurring editing tool enabled scientists to treat DNA and genes as we now use word processing software: to cut, copy, and paste from one section to another. This paragraph began its life in another chapter, but with a couple of simple clicks, I transposed it here, and modified it to make sense. Similarly, the dragline silk gene began its life in a spider, and was transferred into the genome of a nascent goat, tweaked, and positioned so that it makes biological sense. Hundreds of restriction enzymes have now been isolated and characterized, and are cheap and readily available. Even better than merely moving text around, as DNA is a double helix made of two strands that mirror each other with complementary code, there is another useful trick these genome-editing tools can perform. Some simply cleave both strands of the DNA at the same point. These are called blunt ends, and any other blunt-ended DNA can readily stick to it, two ends glued together, like a blank domino. They are perfectly stable once linked, and have value to the gene engineer, as you can stick them in anywhere. But other restriction enzymes cut each strand a few letters apart, creating sticky ends, with one strand of DNA overhanging the other. That makes them an uneven domino, where you can only join to the end if you have the matching piece. A stretch of alien DNA that has the complementary sticky end can join with the host DNA, clicking neatly into place. This means inserts can be oriented in a particular direction, and other useful inserts added to make more complex genetic pathways, a run of dominos built by design.

In the lab almost all of these molecular biology experiments involve transferring minute quantities of colorless liquids from one tiny tube to another. The action of restriction enzymes on DNA is invisible to the naked eye, and we check success with a battery of proxy visualizations, such as measuring the mass of the cleaved or rejoined DNA fragments. It's often far from dramatic, and not easy to see the profoundly unnatural acts being perpetrated. Ultimately, though, the real test is to see whether and how these reconstructed codes work in cells and in organisms. Sometimes the effects might be gross and visible. But frequently, the impact of genetic modification might be subtle or cryptic, and so being able to detect where and when your experiment is working is a fine art.

In genetics, making your transgene, the code with which you are messing, is only the first step. The code does nothing on its own, so it has to be placed into a cell where decoding will occur, and you hope the subverted message will be enacted. In Freckles, the code is not merely the spider-silk gene, but instructions that say when it will be active, or "expressed" in the language of genetics. By careful construction, the silk will only be produced in the milk, even though the gene is present in every one of Freckles' cells.

During the 1990s, extra bits of genetic code added to the tool-shed included colored tags so that we could see while peering through a microscope whether the gene had been successfully taken up into its new carrier. More recently, fluorescent tags have been added so that we can see where the gene is active once it has been integrated into an animal. The most visually striking example of this is the functionally named green fluorescent protein (GFP), which in its natural host, the jellyfish *Aequorea*, glows green in the pitch-dark oceans. The gene that encodes this beacon was extracted from jellyfish in order to be added on to the gene you are testing. Just as in the spider-goats, the machinery that translates the DNA into working proteins is indifferent to its meaning, so reads the universal genetic code without species prejudice, and glows to show that your modified gene is working.[3]

Genes never work in solitude. Instead, they are parts in cascades of networked activity, like a series of conditional instructions. Every cell contains every gene in that organism's genome, regardless of whether it is needed. The choreography of gene activation is therefore of paramount importance. The activation of one gene might trigger the activation or expression of another, or tell it to stop, and by this process we develop from a single fertilized egg cell to a collection of hundreds of coordinated cells, each performing different functions as a result of the

---

3. There is plenty of artisanal skill required to get the additional DNA to function. The syntax and phrasing of the inserted gene have to make sense to the cell's translation machinery. The arrangement of the letters of DNA—*A, T, C,* and *G*—is precise and crucial to the protein that a gene encodes. As discussed earlier, those letters, or more properly bases, are set in threes, each triplet encoding an amino acid, a kind of biological waltz. If your insert comes in on the wrong beat, the dance makes no sense. But when you introduce the jellyfish gene so it starts on the right beat, wherever your experimental gene is active that cell will glow green. This tag is small enough not to affect the function of the protein, just like someone wearing a brightly colored hat while waltzing.

expression of specific sets of genes, rather than an amorphous blob comprising identical cells. This dance is elegantly choreographed, but with limited flexibility. A gene activated at the wrong time, or in the wrong place, or for too long can lead to disease, abnormality, or often fatality. Cancer cells are ones in which genes that would normally tell other genes to stop cell division have failed (or conversely, genes that tell the cell to reproduce are permanently active), so the cell only divides. This uncontrolled reproduction grows into a tumor.

The command to activate a specific gene will typically come in the form of a protein physically attaching itself not directly to the gene itself but to the DNA nearby. These instructions are called regulatory regions, like instructions for flat-pack furniture. One gene will produce a protein that will bind to a particular run of DNA bases near another gene telling it to become active, and a continuous network of activity buzzes inside each cell and between cells to create a life. A large amount of most genomes is not genes—in other words, DNA that encodes a protein. Genomes are littered with these regulatory regions, which don't contain code themselves but act instead as very precise stage directions for the proteins to enact their role.

Through the techniques of DNA modification, turning these networks of instructions into tools has rendered molecular biology an experimental science. As is so often the case in biology, we can determine how things work by breaking them. Traditionally, this has taken the form of studying hereditary diseases or by looking at mutated animals. As mice and men share many genes, we can knock out a gene in a mouse embryo by modifying its DNA to see its result. This is done using restriction enzymes, imported regulatory bits of DNA, and spliced-in transgenes, together forming a process of major reconstructive DNA editing.

Extraordinary though they may sound, these techniques have been the meat and potatoes of genetics for more than a decade, and the industrialization of the process of gene splicing and editing has made genetic experimentation easier. We have taken the tools provided by evolution and modified, designed, redesigned, and honed them to provide service beyond the roles they play as a result of natural selection. A typical experiment to determine the function of a human disease-causing gene would be to isolate it in a family that suffers from that disease, using the same principles of genetics and pedigrees we have known about for a century. Once purified, it gets copied a million times, so there is more DNA to play with. That makes inserting the modified gene into a precise orientation in a DNA envelope much easier, and it can then be posted into a bacteria. A tiny burst of electricity will punch temporary holes in the bacteria's cell membrane; the constructed DNA will just flow in, and with skillful design and a bit of luck it will be incorporated into the host's genome. Every time the cell then divides, the inserted gene will be copied along with the host genome. This is a numbers game, so bacteria are the perfect tool not just because we can manipulate them, but because they grow like crazy. The next step is to destroy the bacteria and leave only your modified DNA. At that point, you can do what you want with it, such as make RNA versions of it, which will show where the gene is active in preserved tissue on a slide, in an organ, or even in a whole animal. Or you can put that gene and all its extra controls into the embryonic stem cell of a mouse, and implant that into a mother to see what it does as the mouse develops.[4]

---

4. This process is called cloning but is unrelated to the popular sense of that word, to reproduce an identical organism as in science fiction or in animals such as Dolly the sheep. My own atomically minuscule contribution to the field of genetics was in a team that worked on a mouse born naturally blind, with shrunken eyes, a deformed retina, and no optic nerve. By various techniques, our team, led by Jane Sowden at the Institute of Child Health in London, excised and spliced various genes and genetic elements to characterize what the faulty gene was doing and when. Using the basics of cloning, we cut the gene out of the genome of the mouse and used it to determine the precise time and place where it should have been working. We tagged it and looked for the other genes it was controlling and being controlled by. By doing this, we helped to discover that the equivalent gene in humans also causes a rare form of blindness in children.

None of that is exceptional, just standard molecular biology going on in thousands of labs all over the world. We have become so adept at using the tools and the language at the core of all living things that manipulating living systems at their most fundamental level is trivially easy. Through experimentation we know how genes in the cells work and, more important, how we can change them. We can correct genes that cause diseases, not yet in humans but in animals, which allows us to study how diseases progress, and therefore how we can treat them. We have developed the ability to read and characterize every single gene in a human, at the time of this writing, in a few weeks at a cost of a few thousand dollars (though both numbers will inevitably continue to fall).

Synthetic biology is the evolved descendant of genetic engineering. It takes the principles of biology and reinvents them with the goal of engineering solutions to specific human problems—disease, environmental issues, and, as we shall see, even space exploration. This synthetic biology movement is a new phenomenon in science, a decade old at the most generous estimate. It has carried with it a subculture ethos, too. But, unsurprisingly, mainstream science and corporations have also begun to take notice of its potential.

Climate change and global warming will define much of synthetic biology's industry and innovation in the next few decades. Expecting people to radically change their behavior is unrealistic, so finding alternatives to fossil fuels is a major challenge. In the meantime, synthesizing fuels, rather than digging them out of the ground, is a partial solution.

There are dozens of projects to make biofuels, converting vegetation into energy in that fundamental way that so many life-forms do naturally.[5] It is a process of carbon fixation, effectively turning car-

---

5. Gasoline itself is a kind of biofuel: crude oil is made of the carcasses of trillions of organisms that have decomposed and been crushed for hundreds of millions of years.

bon dioxide into organic, energy-rich products. In nature the energy of the sun is harnessed in various metabolic processes by a plant so that it can live and grow. Setting fire to a crop will release that energy, now stored in the plants' strong cellulose cell walls. But a more efficient way of tapping into that energy is to ferment the sugars locked in plant cells, also the product of the sun's energy, directly into an oil fuel, which is crammed full of more readily available energy. A handful of companies are attempting to make biofuel from synthetic biology using this very process.

One such project is run by Jay Keasling, a professor at the University of California, Berkeley, who founded Amyris, a synthetic biology company whose aim is to create diesel from live cells. Their designed tool has been a genetic circuit consisting of around a dozen individual chunks of DNA, which they implant into the genome of brewer's yeast, a cell that naturally ferments sugar into alcohol in beer. The success of this circuit in producing diesel is no mean feat, and the scale of their ambition is mesmerizing.

I have been in scores of genetics labs, which range from the charmingly old-fashioned to the ultramodern. Molecular biology doesn't necessarily need the huge clinical white spaces that the movies portray. They are essentially very precise kitchens, with ovens, fridges, and tools for mixing ingredients. But Amyris's labs are stunning. Only the very deep pockets of a commercial venture can provide such luxury, not merely in its space and light, and a healthy San Franciscan café, but in developing the kind of technology that is required in order to industrialize synthetic biology.

One of the most time-consuming tasks in a normal lab is the insertion of the genetic circuit into the host cells. The efficiency of this process is variable. It's not random, but there is an element of hit or miss: the circuit either goes in or it doesn't. In normal lab gene tinkering, we include genetic markers in the circuits, tags that color the cells to indicate whether integration of the alien DNA has been successful. In order to see those colors, the cells have to be grown not in a broth but on a plate, so that the successful colonies of cells can be selected with a toothpick and grown in a small tube, leaving the failures for the trash. It's a trivial but arduous process and it is done by hand. At Amyris, they have mechanized it so that they can pick hundreds of successful clones in minutes, tens of thousands per week, with only minimal human input. A digital camera takes a snapshot of the plate that bears the thousands of yeast colonies and processes it to identify the successful clones. Inside a machine as big as an office photocopier, a bank of needles hovers and zooms over the plate, plucking the successful yeast cells with submillimeter accuracy at the rate of a disco beat.

The cells that have taken up the circuit don't need much more encouragement. Incubated, they simply leak diesel. At Amyris's headquarters in San Francisco they have a three-hundred-liter test tank churning out diesel. The labs smell sickly sweet, like a brewery, but more ciderlike, because the diesel they are refining, farnesene, is found in the oil that gives apples their water-repellent skin. As a fuel, it is cleaner than gasoline-based diesel, as it has no sulfur emissions and reduced nitrous oxides and carbon monoxide.

Creating diesel affords us certain advantages over digging it out of the ground. But it also comes with a different problem. Set aside the fact that clean, synthetic diesel is still diesel, so is still a carbon dioxide–generating fuel. The more significant problem is that the vats of yeast seeping out diesel need to be fed. That means converting plant material, biomass, into food. To grow that biomass is a farming problem, and so comes with the same issues as farming. Amyris has made strong partnerships with companies in Brazil in order to be close to where the first-choice food, sugar cane, is grown. Brazil has the biggest sugar-cane farmers, and since the 1970s Brazil has produced biofuels, predominantly ethanol, as a means of freeing itself from dependence on foreign oil. The key question is this: how much land does it take to make one liter of diesel? The answer is unclear at this point. Different feedstock is digested in different ways, and the synthetic yeast programs can be altered to cope with this. In their Brazilian plants, synthetic farnesene diesel is already being supplied to local vehicles and as an alternative aviation fuel. But Amyris's goal was to grow output to two hundred million liters by 2011 at $2 a gallon. The ambition is clear: Jack Newman, Amyris cofounder and chief scientist, told me, "I'll be excited by a billion liters."

Yet it looks like Newman's billion liters is, for now, a burst bubble. For a while, Amyris looked as if they were going to be the winners in the race to make and sell commercially viable synthetic biofuels. But then the two hundred million liter prediction dropped to fifty million for 2012. In February 2012 the company announced it was scaling back production of farnesene until the industrialization becomes more plausible. The genetic program works just fine, but thus far, scaling it up to the levels where it becomes economically viable does not. For now, for all the stunning laboratories, it seems that the synthetic biofueled future is still a long way down the road.

Meanwhile, we have been able to modify crops to be resistant to blights, grow larger, tolerate frost, and even produce vitamins that protect consumers from diseases. By borrowing genes from a bacteria and a daffodil, scientists have created a form of rice that produces high levels of beta-carotene, a molecule involved in Vitamin A production. Golden rice, as it is called, has the potential to treat the 120 million people worldwide who suffer from Vitamin A deficiency, 2 million of whom die and 500,000 of whom go blind. Yet it remains unavailable due to a combination of scientific and ethical roadblocks.

These stories show some of the promise, potential, and problems with synthetic biology. In essence it is engineering, with the application of basic research at its core. Yet alongside its forefather, genetic engineering, these are immature industries. They face scientific problems, commercialization problems, upscaling problems, ethical problems, and, as we shall see later, dogged resistance from some members of society. Biology is messy, and complex genetic networks underlie that mess. As genetic engineering, just three decades old, gives way to synthetic biology, the great challenge is not merely to simplify those networks, but to commoditize them.

CHAPTER 9

Logic in Life

"If it was so, it might be; and if it were so, it would be; but as it isn't, it ain't. That's logic."

Tweedledee

Flick the switch, the light comes on. This is the simplest useful electrical circuit. The parts are designed and created to follow one single instruction: the switch is a gap when open, but when closed, electrical energy surges through the circuit. The filament in the bulb converts some of the electrical energy into a form that we can detect with the cells in our eyes, and so enlightenment occurs. The function is clear, the instruction pure, the logic is unimpeachable, the light comes on.

At the other end of the scale, you watch a video streamed over the Internet on a laptop. Billions of electrical signals will have been created, modified, and transmitted in order for that moving image to play. The circuits have been designed in intricate detail with each one obeying a logical pattern determined by the hardware and software in your mouse, your computer, the servers that host the file, and so on. The logic is also perfectly clear, but the complexity of the pathway makes it all but inscrutable to almost everyone, and occasionally unpredictable. And yet we use the output of this tangled circuit every day without care or understanding of the millions of decisions that have been made to put a moving image on your screen.

Recall, if you can bear to, some of the electrical circuitry you learned at school. A lightbulb connected to a battery with a simple switch has a binary output: ON or OFF. You might have then added other devices into the circuit to introduce finer levels of control, such as diodes—electrical valves that turn a wire into a one-way street. If you added, for example, a thyristor, you'd have a dimmer switch. Possibly the greatest invention, certainly the most enabling technology, of the twentieth century was the transistor, which allowed the ability to control and modify multiple electrical signals. Interconnected transistors make up what are known as logic gates, which modify the input signal to produce a specific but different output. With the introduction of logic gates, circuits of ever-increasing complexity can be designed and built. For example, an AND gate acts as a positive conjunction: if two electrical inputs feed into an AND gate, both must be ON in order for the output to also be ON (in electronic engineering, capital letters indicate logical calculations, rather than merely bellowing for emphasis). A microwave oven uses this logic. It will only cook if the door is closed and the start button has been pressed. If either of the two crucial signals is negative, then the output is negative.

Electrical circuitry relies on logic, on following pathways determined by the component parts. From the light switch all the way up to the machine I am typing on, the route of information takes the form of digital questions with digital answers: if I press the on switch, the light comes on; if I press the return key (via a long pathway of transistors and thousands of other components), the paragraph ends.

These are the very basics of electrical engineering, a field that has grown from lightbulbs to relaying messages to and from deep space in a little over a hundred years. Every signal sent to your mobile phone, or from the spacecraft *Voyager 1*,[1] is relayed via the logic ingrained in transistors. In a standard microchip, there are billions of transistors, and this is the technology on which almost everyone now relies.

It's a hugely appealing system, not just because it works as we design it to, but because the fact that it works has allowed us to build the modern world. This logic and this ambition are at the heart of synthetic biology. Many components have been built and some have been assembled into basic circuitry. Although the term *synthetic biology* has gained broad usage to include everything from the work of Craig Venter on Synthia to the reinvention of genetic code described later, the field originated and has been maintained by scientists aiming to apply the principles of engineering, specifically electrical engineering, to biology. Living things are immensely complex, with thousands of genes encoding thousands more proteins that interact with one another and the environment to produce millions of cells. But the basic logic of genetics is present in principle: if a gene is activated, the protein it encodes will be activated, and will perform a function.

We think of life-forms not as illogical, yet not with the straightforward formulas of electronics. Nevertheless, the principles of logic gates as they are used in electronics do exist in nature, as complex behaviors and actions often need the parsing of multiple inputs. The carnivorous Venus flytrap plant displays a simple but clever form of logic circuitry when performing its eponymous act. On the inside of the leafy jaws are tiny hairs that act as the triggers for clamping shut.

---

1. *Voyager 1* is the most distant man-made object from Earth, and at the time of writing is eighteen billion kilometers from us. It sends regular messages via Twitter as @nasavoyager.

But that act requires metabolic energy, so it has evolved a mechanism for avoiding fruitless (or flyless) trapping. When a hair is touched by a fly's nudge, a timer begins. If a second hair is triggered within twenty seconds, the jaws snap shut in less than one-tenth of a second. Thus, the input signals that result in the action of trapping are required in conjunction. The double trigger is a form of an AND gate, as both outputs from the hair triggers need to be set to ON for the output of the full circuit to be ON. There's also the timer built into each trigger, so the overall simplified electrical pathway might read: IF TRIGGER 1 + TRIGGER 2 < 20", THEN CLOSE JAWS.

This process is, naturally, managed and run by the mechanics inside the Venus flytrap's cells, with specific proteins that respond to the physical sensation of a fly's touch. But the biological action of a trap waiting to be sprung follows a straightforward logic that looks a lot like electronics: it has no brain or conscious control for the decisions it makes, it just follows a program.[2]

Mostly, however, living things follow a logic that is a far more difficult lock to pick. The activation of genes to produce functional proteins is bound by time and location. Genes interact and respond to the environment in which they are active, in the cell, and from long- and short-range signals from close and distant neighbors. Genes themselves vary from person to person and from creature to creature in small and subtle ways. This can make their function, their output, also vary from person to person. This variation is essential, as it is the basis on which natural selection can act so that evolution unfolds. Combine that with the muddled interplay of genes and environment and we have absolute individuality. This is

---

2. The Venus flytrap circuit is qualitatively different from electronics because the logic of the genes underlying the process is there not to enact the trapping, but to set the trap; that is, to open the jaws and arm the trigger. They express the proteins that act as the components in the logic pathway. In electronics the logic is dictated by the flow of electricity through the circuit.

the reason why fingerprints are unique, even on the fingers of identical twins, whose genes begin their existence identical.

There are some cells that work in a basically digital way. Brain cells—neurons—fire with a dynamic flourish and link up to produce thought and sense. Their spark is only initiated when inputs from other cells (in the form of charged atoms) flow into the neuron until a particular threshold is reached. This process, with a suitably dramatic name—action potential—renders these neurons digital: they are OFF until they are ON. But don't let that simplicity fool you into thinking your brain is easily understandable. There are more than a hundred billion cells in your brain, and each one of them forms thousands of connections with others. This means potentially hundreds of trillions of switches flicking on and off in millisecond bursts every moment you are alive. There may be a logic buried in there, but for the foreseeable future it is inscrutably complex.

For the most part, the switches in life are not quite so binary, either. There are hundreds of other types of cells and switches that are much more nuanced, stimulated on a sliding gauge, or contingent on myriad inputs. Just like a neuron, a gene may indeed be ON or OFF like a lightbulb. But it also may be ON at different strengths, or it may have different functions depending on where or when it is active. This fact goes part of the way toward explaining why we can't understand our own complexity concerning the 22,000 genes present in our genomes (far fewer than had been assumed necessary to describe how humans work). It is also why mutations in the same gene can cause independent diseases in the eye and the kidney. How and when a particular gene is controlled can mark profoundly different functions in entirely different tissues.

There are so many confounding and complicating factors in biology that the underlying logic is lost in the messiness of real cellular life: unpredictable variation set among unfathomable sophistication. A genome, with its full set of biological instructions, is akin to the score of a symphony. The full beauty is expressed in a performance, complete with interpretation and nuance that is not merely encoded in the quavers, crotchets, and minims written on the page.

We sometimes call the seemingly inscrutable complexity of a living system not music but noise. It's a combination of all of the things that we can't quite account for: things such as natural variation in the behaviors of genes, proteins, or molecular interactions that haven't yet been described. The natural progression of biology as a science to synthetic biology has meant confronting the nuances of this mess and scrutinizing it. At the heart of synthetic biology is the desire to avoid that mess by creating new life forms whose circuitry and programming is clear, simple, and, crucially, built not for survival but for purpose.

Sniper Circuits

"Imagine a program, a piece of DNA, that goes into a cell and says, 'If cancer, then make a protein that kills the cancer cell; if not, just go away.' That's a kind of program that we're able to write and implement and test in living cells right now."

Those are the words of Ron Weiss, one of the founders of the modern engineering discipline of synthetic biology and now a professor at the Massachusetts Institute of Technology. He is describing a landmark study his team published in autumn 2011. By using the logic and language of computer circuits combined with biological components, they had built a tool that effectively acted as a cancer assassin.

The terminator circuit is an assembly of DNA components, built to serve a singular mission: to identify and kill a type of cancer cell. Once built, the circuit slots into the genetic code of a virus, itself modified to grant us control over its natural tendencies. When introduced to malignant cells it infects them and, just like all viruses do, adds its synthetic genome (including the assassin program) to the host's DNA. Obliging the natural trickery of virus infections, that host cell unwittingly decodes the killer circuit and enacts the program that will bring about its own downfall.

The assassin circuit represents a set of five questions, each one seeking the presence of a particular molecule unique only to this type of cancer cell. If any of the answers to this molecular interrogation is negative, the program stops, shuts down, and decays, and the cell lives out its normal life. If, however, the answer to all five identification questions is yes, termination ensues. The code in the circuit contains a gene that triggers the host cell's own built-in suicide program. It is simultaneously a conscientious and persuasive assassin, one that makes forensically sure its quarry is indeed the target before quietly asking the victim to kill itself.

The genetic circuit is complex but logical. While it may seem trite to draw such a literal analogy with microcircuitry, this system has in fact specific logic calculations built into it that are entirely derived from electronics. It uses exactly the simplified processing logic of electrical components such as AND gates. It also uses NOT gates, which flip the input from ON to OFF and vice versa.

This particular malignant cell is perhaps the most well-studied laboratory cancer in the world, the HeLa cell. It is an undying ancestry of cells derived from the cervical cancer of a young black woman named Henrietta Lacks. A scrape taken from the cancer on her cervix was grown in a lab in 1951, and was soon identified as being immortal. Normal cells age, grow weary, and ultimately lose their ability to divide. Cell death follows. But HeLa cells reproduce indefinitely due to idiosyncratic faults in their own genetic logic boards. The tenacious immortality of HeLa cells has meant that they are the most studied and scrutinized malignant cells in existence, having been shared and spread among labs throughout the world.

With such a well-studied fingerprint it makes them a worthy testing ground for such a heavily engineered genetic circuit. Most cells are not known in such forensic detail, so foolproof identification would not yet be possible. But for HeLa cells, the five molecular questions allow precise recognition so thorough it qualifies as what is known in electronics as a TRUE value. The precision of the assassin circuit is such that it is in effect a biological computer.

This system is a high-water mark in the nascent field of synthetic biology. As with all potential therapies, it has a long way to go before being available to humans. Currently, it has only been tested in cells in dishes. Animal tests will come next, where the degrees of control will be challenged further by the more chaotic and dynamic noise of a working creature.[3] As a therapy, though, its accuracy is potentially shocking. Cancer, being a multitude of diseases and prone to mutating as it spreads and grows, means that you're aiming at an ever-moving target. Chemotherapy and radiation therapy remain the most effective attack on tumors, though they are damaging both to malignant and healthy cells. Compared with the carpet bombing and collateral damage of radiation therapy, this cancer treatment is a sniper. It represents the immersion that researchers have made in a decade into the realms of programmable biological machinery. As a weapon against a disease, the function of the assassin program is clear, and its assembly is premeditated.

Clocks and Waves

The first forays into the creation of biological components occurred in 2000 with two papers in the journal *Nature* merrily breaching the

---

3. And, indeed, from cell to cell; one of the confounding things about cancers is that their DNA mutates quickly and causes highly variable responses to therapeutic attacks of this precision.

barrier between biology and electronics. One was the creation of a biological clock in the bacteria *E. coli*. By piecing together three sections of DNA that naturally prevent other genes from producing their proteins, Michael Elowitz and Stanislas Leibler at Princeton University constructed a circuit that dictates not merely that a gene is ON or OFF, but is ON and OFF in an oscillating wave. The wave continues as each gene inverts the output of the next: one switches the next OFF, which switches the next ON, et cetera. Cells have a natural cycle as they reproduce and grow, but this circuit was unlinked to that. The output of the circuit was the expression of the gene for green fluorescent protein (GFP), and so the cells would glow green in slow waves.

The second component part was made by a team from Boston University led by Timothy Gardner. They built a genetic version of a type of electric component known as a bistable, or more evocatively, a flip-flop. These types of switches flip between two states, but each of these positions has a function. You push the button on a CD player to turn it on, and again to turn it off, but in the off position, the player is actually on standby, rather than cut off from the main supply. It has two states, ON and ON STANDBY, and flipping between the two is fully reversible.

Both these engineering projects adopted not only the language of life to build unnatural gadgets, but also the language of electronics to convey the functional design ethos these gadgets were bringing to biology. They were given names that could be straight out of the lexicon of electronics (or possibly science fiction): the glowing wave circuit gained a new electronics-sounding moniker, the repressilator; the bistable flip-flop device borrowed from an existing electrical part, a toggle switch.

These two inventions are rightly seen as the first synthetic biology parts: microscopic tools designed to enact a program but built from DNA. The synthetic biology workshop opened with these first parts built, and what followed was a cascade of other mechanisms, tools, parts, and pieces, all made out of DNA, taken, modified, fused, and redesigned from evolution's toolbox. In the last ten years we have gone from the species-barrier-breaking insertions of genetic engineering to an ever-expanding toolbox packed with widgets. There are now switches, pulse generators, timers, oscillators, counters, and logic calculators. The combinations of these and many other components have extended our control of living systems to genes, protein function, cell growth and reproduction, metabolism, and the ways cells talk to one another.

Less obviously practical than the cancer assassin program, but no less complicated, was a circuit built in 2009 by Jeff Hasty and his colleagues at the University of California. Timekeeping cycles are fundamental to most living things. They are known as circadian rhythms and determine all sorts of patterns of behavior in relation to the flow of time through our lives, such as the passing of day and night. We rely on cells to regulate our own metabolisms, such as in the rhythmic release of insulin or in the orchestration of sleep cycles, both of which are disrupted in people who regularly work night shifts. How these timers work is only partially understood, but they use similar mechanisms to the artificial biological clocks built in these synthetic circuits. Hasty wanted to design synthetic circuits that created a regular pulse of activity, like the ticking of a clock. These so-called oscillators work as cellular timekeepers, determined by the careful design of a loop of genetic activity within a bacteria, three genes following a circular set of instructions. Imagine three friends standing on the corners of a triangle, sixty feet apart, all facing inward. Each has a chair but only stands if the friend to his or her right is seated. When the first stands, the second sits and the third stands. The first then sits, the second stands, and so on. This continuous response is called a negative feedback loop, and the reaction time of the three creates a clock of activity. It won't work with four people, as the flow will end as the fourth person will be seated after one round, and that will not prompt the first to sit down again.

As it is, this circuit will continue as long as the people can manage it. There is no output, though. Now imagine that, as well as standing and sitting, the first person alone also has to sing a note, but only while standing. As long as the game continues along those rules, he or she will be singing every other complete circuit. If you close your eyes and only observe the output, you would hear a regular burst of song. This is a basic biological oscillator.

In a synthetic biology oscillator, when gene A is active, it shuts down gene B, which activates gene C, which completes the circuit by shutting down gene A. The program is constructed from those three component parts—genes and their activation sequences—and imported into the guts of a bacteria. In the version involving people, the frequency of the burst of song is determined by the speed at which each of the friends reacts to his or her neighbor. In the cell, it is the speed at which the protein is produced. Gene A is translated into a working protein, which binds to the ON/OFF switch for Gene B, and so on, but each step takes time. The clock function is a result of this delay. As with the note being sung, the output of the circuit in bacteria might be the expression of a fluorescent protein, which will slowly come on and off with a regular pulse. In synthetic biology, the aim is to elicit control from these DNA circuits, so the ability to fine-tune is desirable. Adding other loops and other mechanisms can introduce further control to the frequency and strength of the output, adding to the complexity of the logic.

One limitation of this basic biological clock is that the simple oscillator circuit is self-contained. In clock terms, this would be as if all wristwatches kept time perfectly, but everyone's was set to a different time. Time is most useful as a commodity only if we all synchronize our watches; otherwise, watching *The Ten O'Clock News* would be a lottery. In bacteria, engineering a single cell to produce a regular pulse is one thing, but to turn that pulse into a synchronized wave in a population of bacteria is entirely more difficult. *E. coli* reproduce every twenty minutes or so; it's a busy crowd to exert military control over.

In Jeff Hasty's program, though, the circuit also sends out a signal to all neighbors, which syncs the beginning of the loop. Rather than three people in a triangle, it becomes a wave in a football stadium. You still stand when your neighbor sits, but it happens in banks of people at the same time. The pulse, the output of a hearty cheer, is bellowed not by one person but by thousands. The bacteria, speeded up in time-lapse video, pulse with glowing green, like a luminous ripple. Anyone who has ever been part of a decent wave will know that it's impressive enough. But the size of these bacterial populations is far greater than any stadium capacity. The act of inducing perfect synchronicity in these synthetic bacteria is the equivalent of making every traffic light in the world turn green at exactly the same time.

That's a neat trick indeed, and shows the level of control these new synthetic circuits give us. But it also has valuable potential as a tool for synchronizing the output of a population of synthetic cells.

Insulin's job is to tell the body how and when to process carbohydrates and fats after you've eaten some roast potatoes. Glucose is the fuel that powers cells (and by extension, organisms), but the concentration of this simple sugar in your blood is a delicate balance: too much or too little is deadly poisonous. We take in glucose directly in our diets, and other foodstuffs get converted to glucose in our muscles and liver where it is stored in other forms such as fat. We have evolved all sorts of mechanisms and biological cheat sheets to make sure our cells are fueled by a constant supply of glucose, regardless of whether you have just scoffed a chocolate bar or not eaten for hours. Insulin is an integral hormone to maintaining glucose levels and it does so by instructing cells in various tissues to collect glucose from the bloodstream. In doing so, glucose levels in the blood are reduced, and the production of glucose from stored fat is turned off. The production of insulin is prompted by high blood glucose, and shuts down when it reaches a threshold, so is in itself part of a feedback loop. Importantly, the curiousness of insulin production in the body is that, even when we are at rest, the level of insulin is not still. It oscillates gently at a rate of between three to six minutes regardless of blood glucose, like an idling car. When this process malfunctions, diabetes is the result. A hypothetical, synthetic circuit with the ability to oscillate in its production of insulin (or anything) in a whole population of cells, rather than in an individual one, would be of enormous benefit in mimicking the natural pulse and ultimately be a potential therapy for diabetes patients.

As with Ron Weiss's cancer assassin program, these engineered biological circuits are not nearly ready for clinical use. That program is carried in a modified virus, unable to reproduce and unable to enact its software without the machinery of existing living cells. The virus has to infect the cell before it can enact its diagnosis and sentence. Most of the circuits of synthetic biology are integrated into the genomes of bacteria, but as of now we are unable to test synthetic circuitry in humans. A long time before we reach that point, there is the not insignificant problem of delivering the software and its packaging to the place where it's needed. The bacteria themselves are merely vessels that carry the software and provide the mechanics of producing that circuit's output. Humans are covered in, and full of, bacteria, which are much smaller than human cells, and so we can carry maybe ten times more bacterial cells than our own. Almost all are benign or beneficial, especially in our guts, where billions of bacteria—our microbiome—provide digestive functions we don't inherently possess. However, our bodies are good at detecting and destroying invasive species, so genetically enhanced bacteria would need to be undetectable to the immune system, coated in a molecular invisibility cloak in such a way that the legions of intruder alarm cells patrolling our bodies simply fail to identify the synthetic cell. Yet our immune systems have had billions of years to evolve the ability to detect foreign invaders, and free-roaming synthetic bacteria would be policed harshly. An alternative could be to package the bacteria and hide them from the immune system. Unlikely though it sounds, NASA is working on precisely that problem.

Under the Skin

NASA's base at Ames resembles a archetypal small American town, just off at Californian freeway, with wide roads, grassy squares, faux stucco buildings, and acres of blue, blue sky. Except that in between these blocks of normality, there are colossal paeans to the grandstanding 1950s optimism that science would deliver the future. The biggest wind tunnel in the world occupies several blocks, its interior roomy enough to accommodate a full-size airplane. The air intake is a giant black rectangular grille so large that it's difficult to gauge its colossal scale, until I spy some NASA scientists playing street hockey, dwarfed at its base. Four blocks to the north is a gargantuan hangar built in the 1930s to house the building of dirigibles, short-lived rigid airships. In a clean room (i.e., a room free of dust and dirt) around the corner, the next Moon-bound spaceship, LADEE, hovers as its carbon dioxide–based levitating propulsion engine is tested by technicians.

Naturally, NASA exists to explore space. On the corner of a spacious crossroads there is a building whose concrete facade is pockmarked with casts from meteorite impacts. Inside, the researchers in NASA's synthetic biology program are using the microscopically small to solve some of the biggest problems that lie between us and our desire to explore strange new worlds.

The sun beats down at Ames; while being the bringer of energy and life on Earth, it is far less charitable off-world. Sporadically and unpredictably, our star will blast out a shot of extremely high-energy particles with enough clout to disrupt power supplies on Earth. Similarly, other stars in the universe eject a continuous bombardment of high-energy particles from deep space that bathes the galaxy in radiation. Beyond the protective blanket of our atmosphere solar flares and galactic cosmic rays combine to be one of the most restrictive dangers for human space exploration. Radiation, including in these interstellar forms, has the effect of slicing up DNA randomly. Often that is not a problem: DNA repair is a major industry within our cells, with legions of copyediting proteins beavering away to make sure that errors in the code, or mismatches in the two strands of the double helix, are identified and patched up. Occasionally, the damage may occur in one of the many genes that control cell division; if that happens in a gene whose purpose is to instruct a cell to stop duplicating itself, then you have the beginnings of a cancer—uncontrolled and unrestricted cellular growth. Conversely, breaking up the many genes necessary for the cell's continued existence can result in the inception of a natural cell suicide program.[4] Unrestrained cell death or cell growth is equally bad for an organism's well-being. We are constantly exposed to damaging radiation, but mostly in small doses or infrequently. This is why, when you have an X-ray in a hospital, the technician stands behind a protective screen. It's a rare event for you, but unshielded, technicians are exposed dozens of times every day, with potentially deadly consequences. In outer space, the exposure is continuous.

In 2005, the U.S. Federal Aviation Administration set off on a

---

4. Cell death is, perversely, essential for growth and a very important part of development in the womb. For example, as fetuses our hands start out as paddles, and the fingers grow within them. The fingers separate when the cells in the webbing enact their own death.

virtual round-trip to Mars, including a fourteen-month stopover on the red planet. They calculated the amount of exposure to space radiation astronauts would receive, and deduced that astronauts would suffer a significantly increased lifetime risk of cancer (as well as a host of other conditions, including cataracts and sterility) as a result of exposure to cosmic rays and a bout of radiation from a solar flare. The safety limit was a journey of around fifty million miles—just short of halfway to the sun. Shielding on the spacecraft prevents radiation exposure to humans, but it is heavy, and weight is a critical issue when calculating the propulsion and cost of lifting off of our planet. The hostility of space means that our biology is just as limiting as our engineering when it comes to exploring the heavens.

Our immune systems do have built-in defenses that counter the effects of radiation damage. We have these mechanisms in place to deal with the much weaker radiation—ultraviolet—that causes sunburn and other forms of DNA damage. When detected, the severed DNA switches on genes that produce small molecules called cytokines, which flit from cell to cell, triggering cascades of repair programs. But they have limited reach, especially in response to prolonged exposure to damaging radiation, and this is why cytokine therapies are sometimes used to enhance the body's natural response.

David Loftus and his team at NASA Ames are planning a designed synthetic program in which the input is a detection system for radiation or DNA damage, and the output is the controlled release of cytokines. This program is different from the cancer assassin in that the output is a molecule that stimulates the body's own natural immune system. Carrying the program in a virus that will infect a cell will not work; the production needs to be self-contained. Therefore, a bacteria is the most appropriate synthetic cellular manufactory. The question is how to introduce a bacteria safely into humans without provoking an immune response.

The circuit is not yet built, and this construction will face all of the same problems as any synthetic biology project. But Loftus and his team have made stunning advances in the construction of the capsule to contain the cells. As if the circuitry of synthetic biology was not impressive enough, the biocapsule is made from carbon nanofibers. A minuscule mold sits on top of a hypodermic needle, which is attached to a pump. This is submerged into a suspension containing the carbon nanofibers; they are literally sucked onto the mold to form a capsule. This tiny pellet is half a centimeter long and half a millimeter wide, not much bigger than this letter I. Although the capsule is tiny, bacteria are much smaller, so it is large enough to imprison tens of thousands of them.

It is also biologically neutral, which means the carbon nanofibers will not provoke an immune response and will cause no harm to the astronaut. But with serendipitous grace they bunch together in a porous tangled mesh, like a sheaf of frozen worms, visible only under electron microscopy. The gaps running through this mesh are too small to allow bacteria through, but big enough to let small molecules, such as the cytokines that tackle radiation damage, seep out. The idea is simple enough. The synthetic bacteria cells are placed within the capsule, and the capsule is implanted under the skin of the astronauts. The cells contain a program that produces and releases cytokines upon exposure to solar and cosmic radiation. It requires no diagnosis or intervention, and, elegantly, the treatment for the sickness is prompted by its cause.

Imagine this technology applied to less exotic practical human problems: a synthetic cellular circuit permanently implanted under the skin to deliver the treatment to a disease without the patient ever knowing. The potential is breathtaking. Diabetes sufferers stand out as being obvious benefactors. The cells in the pancreas that produce insulin could effectively be replaced with synthetic organisms that serve exactly the same purpose. Instead of injecting insulin, a synthetic circuit borne by bacteria hidden in a biocapsule shell would produce insulin according to the body's fluctuating needs. In principle, the patient would never even be aware that this process was going on.

As ever with brand-new, untested inventions, this exact mechanism may yet come to nothing. It may turn out to be impractical or too expensive to be of general use. At best, this system is a decade away: the capsule is still being developed, and the synthetic circuits themselves are years from being tested in animals, therefore years from being tested in humans. Yet this shows how the need for practical solutions to complex problems typifies the renaissance nature of synthetic biology.

Reality Check

What I've described above is the promise and the hope of this new engineering project. I've outlined some of the highlights of ambition and success of this emerging endeavor. Currently, there aren't that many more, though the volume of synthetic biology publications is rapidly increasing. It is easy to get swept up in the hype as the rigor of the lab is translated into terms aimed at nonexperts designed to inform, enthuse, or generate criticism. However, the comparison with electrical engineering is one that has been made by synthetic biologists themselves, in an attempt to standardize the parts of DNA into component parts in order to construct living machines.

Yet the reality is that referring to living things as machines glosses over the facts of life. Cells and organisms are machines whose complexity is many orders of magnitude greater than an engine, a production line, or even a computer. Engineering, like evolution, is an iterative process. The incomprehensibly deep time since the origin of life on Earth has made the number of tests, failures, rebuilds, and further tests undergone by the mechanisms of life an impossibly large number. Every organism that has ever existed was an incarnation in a perpetual test of engineering function, unguided and ruthless: will it live, and live to reproduce itself? This is true, naturally, of any species or organism, but the real selection under scrutiny in evolution is individual genes, the functional natural component parts of living things. These work in consort, in networks, in cascades, and in meandering dynamic pathways. The complexity of a living organism has billions of years of successful testing behind it, and that means that to simplify or design genetic circuits into simple (or even complicated) tools is tricky. Biotechnology expert Rob Carlson told *Nature* in 2009 that "there are very few molecular operations that you understand in the way that you understand a wrench or a screwdriver or a transistor."

The test of design in engineering is "Does it work?" and, more precisely, "Does it work as we designed it to?" In this new discipline the answer for many of the several thousand parts and circuits currently available is a simple no. The complexity of biology in cells generates noise—unpredictable variation that masks or subverts the intended output. The so-called repressilator that marked the beginning of synthetic biology in 2001 by making bacteria blink glowing green worked impressively well. But the surprise was that not all cells worked the same. The pulse of the glow was nowhere near consistent: some cells were brighter than others, some were slower, and some skipped a beat. The reasons why are not impossible to understand, but complexity breeds inscrutability, and that makes the reductionist ethos of engineering harder to fulfill. Once again, the analogy with electrical engineering even holds true here. We already make digital logic boards so dense and so filled with software that we don't fully understand their behaviors. This is why computers crash. These systems are built for a purpose and are tested and retested to make sure they succeed at the tasks we design them for.

Still, that does not mean that we know or predict all potential behaviors that they may express. In computer hardware design, failure analysis is critical, just as it has been in human genetics for a century. We have always discovered genes and their functions only after observing them fail, that is, when they cause disease. Well-designed hardware is built to cope with the unpredicted and unpredictable noise that inscrutable complexity may host.

The problems caused by mutated genes that result in disease have a clear origin if you know where and how to look for it. In the wishfully simplified programs of synthetic biology, it is the unforced errors that are problematic. Decoding the obvious faulty mechanisms of diseases and designing successful programs are certainly the aims of genetics and synthetic biology. Yet understanding noise and catering to it remains a puzzling obstacle. Because of this, the parts, the circuits, and the living hosts make the dream of a standardized DNA construction kit an unrealized one as of now. However, this is a young field, immature and full of hope. Typically insightful, Isaac Asimov, the doyen of twentieth-century science fiction, noted that the best moments in science are not when you exclaim "Eureka!" but when you say, "Hmmm, that's interesting." That noise, the fact that these programs don't quite work as designed, is definitely a problem—but a most interesting one.

Nevertheless, parallels with computing are striking. The legends of the computer industry's roots describe future billionaires like Steve Jobs and Bill Gates tinkering with electronics in their garages, playing with components and code to make better hardware and better software. The results were Apple and Microsoft, a computer on every desktop, the Internet, Google, and a smartphone in every pocket.

Whereas political revolutions tend to be defined by events, such as the storming of the Bastille or the execution of a head of state, the cultural revolution of computing was not triggered by any single act.

Similarly, workers in nineteenth-century Europe were aware that newfangled technologies such as looms and spinning jennies were driving significant changes to their working lives, and those changes were happening at a fast pace. But the aggregation of that change and its significance were unknown at the time, and the term *industrial revolution* was introduced much later. The practitioners of synthetic biology are creating in a way that is novel for biology and for science, all within one decade. This manipulation of four billion years of evolution, specifically for the creation of unnatural biological tools, is potentially a revolution, one that is occurring right now.

Richard Feynman's aphorism is crucial: to understand something you must first be able to build it. The terms, the language used, and the techniques are all borrowed from the problem-solving ethos of engineers, a reductionist view of function through design. The extension of natural processes to serve humankind's desires is not new, but has rarely been approached with such a clear sense of built-in construction. Though the founders of synthetic biology had electrical engineering in mind, the metaphor that has stuck most is a well-known toy. Just like DNA, a Lego brick is universally adaptable, designed to fit with other pieces from other sets. Similarly, synthetic biologists have tried to make the individual components of their circuits interchangeable, so that construction of novel circuits is not hindered by the natural noise of biology. With that in mind, the spirit of invention, engineering, and creativity have been built into synthetic biology in a unique way. Yet as this revolution unfolds, questions of ownership, legality, and ethics press urgently upon us.

CHAPTER 10

Remix and Revolution

"Rien ne se perd, rien ne se crée, tout se transforme."

Antoine-Laurent Lavoisier

E volution is the most creative enterprise that has ever existed. Nothing comes close to the diversity, sophistication, and beauty that have transpired at the hands of DNA and natural selection. At its heart are two words that carry a potential derogatory sense: *imperfect copying*. In another sense, evolution is the ultimate exemplar of that maxim made famous by Isaac Newton—"If I have seen further it was by standing on the shoulders of giants"—which neatly describes the derivative nature of creativity while exemplifying it, as he was borrowing and adapting the same idea from twelfth-century philosopher Bernard of Chartres, who was in turn probably referring to ancient Greek versions of the same idea. The French chemist Antoine-Laurent Lavoisier's version in the epigraph above was a comment on the nature of matter, but works equally well for energy, biology, and ideas: "Nothing is lost, nothing is created, everything is transformed."

It's easy to apply a similar principle to culture, that while there may be nothing completely new under the sun, the creation of new ideas is still born of copying, adapting, and transforming what has come before. In music, for example, it's not too difficult to trace an evolutionary path of influence through several centuries of creativity, from Bach to Haydn to Mozart to Beethoven to Mendelssohn (or, in modern times, from Chuck Berry to the Beatles to Bowie to punk to Joy Division to the Smiths to the Stone Roses and on and on).

The nature of copying in music changed in the 1960s when a new method of musical creation emerged that was unlike anything that had come before. Technology enabled musicians not merely to copy and modify what had come before, but to directly borrow, co-opt, and lift it. The sampler allowed music producers to take the drums from one existing track, the horn section from another track, and the vocals from a third and mix them together with any other elements to create a new sound.[1] As a technique for creating new sounds, sampling really took off on the streets of New York with the emerging hip-hop scene (then referred to as rap) of the 1970s. DJs in the Bronx would simultaneously entertain crowds, pay homage to their musical forefathers, and create radical new sounds by mixing records together on twin turntables. They would take the beat or riff from one record and drop rhythm or lyrics on top from another, often using soul records as their licks. In live performances at clubs, and later in recordings, they were not playing new music but creating it by adapting and remixing existing sounds. By the early 1980s, sampling machines were being used to record short extracts from a record so that they could be integrated into a new tune. It could be slowed down, speeded up, looped, or stretched to help create a brand-new sound that was derived from and built on the shoulders of previous works.

Synthetic biology is remixing. The ethos of this emergent scene

---

1. The Beatles did this in 1967 by borrowing and looping calliope recordings on "Being for the Benefit of Mr. Kite!" on *Sgt. Pepper's Lonely Hearts Club Band* (and even more so on such experimental tracks as "Revolution 9" on *The White Album*).

is one of unprecedented and unbridled creativity that is a remix culture. DJs did not have to be virtuoso players of specific instruments in order to create new sounds. In synthetic biology, the creators do not need to be geneticists, DNA technicians, or even biologists to build new organisms. The principle is simply to create.

Since inception many of the pioneers of synthetic biology have strived to build a movement of open, democratic science, where there is a free exchange of ideas, techniques, and materials. If year zero of this new industrial revolution was the repressilator and the toggle switch in 2001, the first big inroads were made a couple of years later. The transition from tinkering in labs to a movement in science occurred during 2003 and 2004, with the emergence of two phenomena that became emblematic of the spirit of synthetic biology.

One was the founding of the Registry of Standard Biological Parts. For all the talk about the wonder of the interchangeable genetic alphabet, it takes a lot of code to make a life-form. That means there is a lot to play around with, and more important, to get wrong when remixing your code. Ironically, for something to be open and free for everyone to use, it needs to be codified and standardized. Think how annoying it is that electrical plugs are different in different countries, or phone chargers change from one model to the next. The language of music is universal, but you need a whole battery of adapters and plugs to perform it in different countries. Yet nuts and bolts are standardized, so that thread size or screw dimensions don't have to be reinvented every time you want to assemble a piece of furniture.

Transfer that to the unlimited amount of coded information contained in the new circuits of synthetic biological components. Dozens, and then hundreds, of labs started building their own gadgets and devices, all of which only work in circuits built into genomes of living things (or, in some cases, near-living viruses). Without standardization, useful sharing is hamstrung. Thus, in 2003, the BioBrick project was conceived by Drew Endy at Stanford,

Tom Knight (then at MIT), and Christopher Voigt at University of California, San Francisco (UCSF). The BioBrick part is a useful and simplified way of plugging in bits and components so that everyone who signs up doesn't have to redesign each idea and each part for him- or herself. BioBrick parts are to genetics what the sampler is to music— a system for freeing up the elements of the rich biological past and the ingenuity of others' designs in order to foster extreme forms of creativity by standardizing the assembly of DNA components so that they don't need to be redesigned every time. This is where the comparison with Lego holds some weight. Lego bricks are beautifully designed to connect with any other Lego brick; there is a universal fit to the pieces, and they can be assembled regardless of their original kit. At the time of this writing, the BioBrick catalog contains more than ten thousand parts. Each one is a piece of DNA and is delivered in the mail as a dot on some blotting paper. Drop it into a solution, and the DNA floats off the paper, ready for assembly. Some BioBrick parts are genes, some are regulatory instructions, and some are combinations of both, already assembled to be integrated into new circuits. All have been standardized so they can be knitted together for new creations.

The other transformation that occurred at this time was the inception of the International Genetically Engineered Machine (iGEM) competition in 2003 and 2004. Each year, teams of undergraduate students devise a problem and design and build their proposed solution using only the components available in the Registry of Standard Biological Parts. Students compete to get into their college teams and spend weeks researching, designing, and enacting their solutions to their chosen problems. It is a talent competition for the brainy, and the excitement is palpable. Each team receives a kit consisting of a thousand BioBricks from the registry.

After rounds of regional selection, the finalists meet up at MIT in November for the annual jamboree. The teams show off and celebrate their creations, and synthetic biology pioneers and leaders decide a range of prizes. The grand prize is emblematic of the construction mindset: an oversize aluminum Lego brick.[2]

The democratic principle is built into the registry and the iGEM competition, too. The resources are all free, but the expectation is that all players will contribute. The registry website declares that users "get some, give some." All teams compile their notes and projects, successes and failures on open wiki pages designed to encourage sharing and standing on the shoulders of peers.

Although few of the iGEM projects have come to fruition yet, there is a kind of vibrancy about this endeavor that is unlike any science meeting I have seen before. Many undergraduate students do summer projects in their tutors' laboratories, where they often do menial or slavish jobs, part of the grind of data production that science is based on, and in so doing gain experience of the reality of lab life.[3] At the same time, iGEM teams create new and potentially important projects and solutions with a zest that is thrilling. They spend their summers inventing solutions to global problems. They are combining brand-new, often untested tools, and in so doing are building new ones using the most advanced biotechnology available. They then present their work to the judging panel, which is made up of a hefty proportion of the most significant synthetic biologists working today.

Hello World

In 2004, a striking entry came from a team-up between the University of Texas at Austin and UCSF. They designed a circuit and

2. This was replaced for the 2012 competition by an elegant wooden giant Lego brick.

3. In the summer of my second undergraduate year in 1995, I spent three months measuring the vital statistics of three thousand stalk-eyed flies to assess whether asymmetry was greater in their eyestalks compared with their wings. It was not. The world of biology remains utterly unmoved by this study, but the lure of the lab was strong enough for me to return.

successfully integrated it into bacteria, which then acted as a photographic plate. One component was a light-sensitive protein, which would perform its function when stimulated by a beam of light. They linked this protein to a standard pigment-producing gene called lacZ. When they grew the synthetic bacteria on a flat clear plate and shone an image onto it, the bacteria processed the light and the dark at the resolution of one hundred megapixels per square inch. The work was published in the journal *Nature*, and the components developed into a new synthetic biology device: an edge detector. The projected image was a text message, which appropriately enough announced the arrival of this new field, but was also an in-joke displaying synthetic biology's computing roots. The phrase is the output of a standard test that programmers use to check that their coding language is functioning correctly. It simply read, "Hello World."

The grand prize winners in 2009 were from Cambridge University: they extended the use of green fluorescent proteins (GFP) and designed a system for bacteria to produce multicolored dyes as a tunable detection tool. These are known as biosensors; they already exist for such purposes as detecting blood glucose levels. However, every biosensor so far has had to be designed from scratch and built specifically to detect just one target. The Cambridge iGEM team aimed to build a generic biosensor that could be tailored to multiple uses. For example, the engineered bacteria could be tuned and used as a detector of a specific toxic chemical in the environment, and produce a color-coded pigment in response to its concentration. They also eschewed the use of GFP, as this requires a fluorescence kit to see it; their aim was to make a biosensor that could be seen with the naked eye. The parts of the circuit are the sensor, a sensitivity tuner, and the pigment generator, six of which are new parts, and were submitted back into the registry for others to use.

Washington University took the grand prize in 2011 with a circuit that produces forms of fossil fuels. The runners-up, from Imperial College London, attempted to fight desertification—the gradual erosion of fertile ground into agriculturally and therefore economically useless land. By 2025, some estimates predict that as much as two-thirds of Africa's arable land will be rendered barren by this phenomenon. The Imperial team has engineered genetic circuits in cells that will nurture roots, which in turn shields fertile topsoil from the eroding elements. Seeds are coated in the synthetic bacteria, and once they germinate, the cells inveigle their way into the infant roots; the program they bear produces the plant hormone auxin, which accelerates growth into the topsoil.

The Slovenian national team appears to win the iGEM grand prize in even-numbered years, claiming the prize brick in 2006, 2008, and most recently in 2010, when they eschewed the language of DNA altogether. Instead of manipulating the code embedded in the A, T, C, and G bases of DNA to create a new function in translation, they used the simple fact that we can put those letters in a specific order to build a miniature production line. Living things are full of such production lines: one gene produces a protein that triggers another reaction, which triggers another gene, and so on. Unlike the ordered machinery of a factory, however, the inside of a cell is a busy milieu, with all the code and proteins effectively free-floating in the gooey plasma. Real factories would be disastrously unproductive if each part was allowed to randomly drift to the next step on the production line. Slovenia's team designed a sequence of DNA that organized the unsecured mess into a neat line by using DNA as a production-line scaffold. Instead of having proteins free-floating in the cell, they figured that they could tether them to a unique sequence of DNA. Many proteins bind to DNA anyway, so much is known about how to get a protein to clamp onto a specific sequence of DNA. With the right stretch of DNA each protein in a pathway can be pulled out of the cellular soup and tethered sequentially, resulting in a biological production line. The primary aim is that this scaffold will radically increase the efficiency of synthetic biology pathways. It also opens up the possibility of more tightly controlled oscillator switches or other standardized devices in the synthetic biology tool kit.

Careful Packing

Almost all of the iGEM teams come from universities. As described in the previous chapter, though, NASA has been quick to identify synthetic biology as a discipline that will enable its continuing mission: to boldly go. One of the more fanciful ideas NASA is exploring is synthetic bacteria that make bricks.

This challenge was taken up by students in the iGEM competition in 2011. An hour from San Francisco, Stanford University collaborated with nearby NASA Ames as well as Brown University to figure out how synthetic biology could help us build colonies on other planets. Led by Lynn Rothschild, they have asked how synthetic circuits can contribute to terraforming. As mentioned earlier, the issue of weight is absolutely critical in space travel. It is estimated that the act of launching a thing beyond the earth's gravitational pull costs around $5,000 per pound. The mass of *Saturn V*, carrying all the modules to get the three crew members of Apollo 11 to the moon, enable them to have a stroll for a few hours, and come back home, was around 3,000 tons. The lunar module itself, the *Eagle* that landed, was 17 tons. Estimates of a round-trip to Mars include consumables and life support of 200 tons alone. Much like packing for vacations, minimizing what you take with you and using what you find when you get there is the best way to do it. In situ resource utilization (ISRU) is Nasa's official name for this method, but I prefer to call it careful packing.

Given that our dreams of establishing permanent bases on other worlds are profoundly hindered by having to carry the building materials from Earth, using the materials on another planet is appealing. The question is, can we build with moon dust?[4]

The Brown-Stanford iGEM team founded their project on this question. They happened upon the alkali-loving bacteria *Sporosarcina pasteurii,* whose metabolism involves expelling the mineral calcite as a waste product. In the right type of sand this undergoes a process called biocementation, that is, it forms cement. Water in normal concrete forms a sort of glue by a range of chemical reactions and is used up in the process. The cells that secrete calcite can serve this purpose without water, other than what they need to live. The key component of this process is a protein called urease, which initiates the breakdown of urea, and in these bacteria ends with cementation outside the cell. The iGEM team identified the genetic circuit for urease production in *Sporosarcina* and copied it into a standard lab cell, *E. coli.* They then characterized and standardized it into a BioBrick part, where it became known as Part: BBa_ K656013. When expressed in a cell, and placed in regolith, the circuit is activated and has the ability to cement it into bricks. They

---

4. In the 1980s, a team of lunar mechanics got to play with a huge amount of regolith, the powdery dust that makes up the moon's surface. A huge amount, in this case, is 40 g, about the equivalent of four tablespoons of sugar. This precious sample, brought home by the astronauts of Apollo 16, was cemented into tiny bricks, an inch-by-inch cube and a couple of other shapes, and was tested to destruction. The researchers concluded that "Lunar soil can be used as a fine aggregate for concrete." One of the key ingredients of concrete, however, is water, and though the moon is not entirely desiccated, there isn't a lot of water there. In 2009, LCROSS was deliberately crashed into the Cabeus crater, a pit shrouded in permanent darkness on the south side of the moon. NASA very carefully watched the mile-long plume that erupted after impact and saw from its unique ultraviolet signature that water and ice puffed out. It's buried, though, and not available in quantities that would enable construction of a moon base.

have tested this process using simulated Mars sand, admittedly in molds less than half an inch across, but within a few hours they solidify into tiny bricks. Clearly, this is an astronomical distance from being used as a building material on Mars itself; that journey is at least a decade away, but at the rate synthetic biology is progressing, making actual bricks from Martian regolith becomes a whole lot more plausible. True to the playful roots of the BioBricks project's mentality, they are called REGObricks. The weight saving is enormous. A small vial of these remixed cells could be transported aboard a ship weighing only a couple of grams, then grown into a REGObrick factory on the planet itself.

These are just a handful of some of the hundreds of entries to the iGEM competition during its first decade of existence. The Bio-Bricks Foundation aims not just to spread the potential of applying engineering principles to biology, but to do it in a way that fosters open, egalitarian science, creativity, and a free exchange of creativity, much like the remix musicians in the 1970s and '80s.

While this may sound somewhat idyllic, there are problems with iGEM and the BioBricks Foundation. In terms of bearing fruit, more often than not, iGEM teams do not actually deliver the products they have designed, often due to the time constraints of the competition (though many individual parts and circuits have resulted in published scientific papers further down the line). But there are plenty of other issues with the BioBricks Foundation. Hundreds, if not thousands of the parts in the registry are not well characterized. The individual parts include genes as well as the instructions to turn genes on or off. But many of these behave inconsistently or unpredictably, and much of this may be to due to the biological noise of a living cell, as compared with the logic diagram on a design. The nature of the competition means that students often don't have time to fully characterize the parts they have built. Standardization is becoming a Sisyphean task, with the rate of submissions overwhelming the urgent task of characterizing what is already there.

Nevertheless, the excitement of the competition is electric. Teams contain students from backgrounds in math and engineering as well as from the biological sciences, and bring a fresh ignorance to the problem solving. I went to the UK team meet-up in September 2012, organized by the University of East Anglia's team leader, Richard Kelwick. The UK entrants presented their projects to one another and a handful of synthetic biologists in anticipation of heading off to Amsterdam to be selected for the world finals. Out of eighty-five students entering from the United Kingdom, twenty-nine were not biologists. The entry level for some of the most sophisticated, creative, and advanced genetic engineering is as open as it has ever been. Of the three teams that went through to the next round, University College London's project was the most outlandish, literally: bacteria that would collect and consume the billions of tiny fragments of non-biodegraded discarded plastic that float in the oceans, with the intention of using this harvest to build a plastic island, recycled and claimed from the sea.[5]

Open access is deliberate and inherent in the BioBricks mentality: a profound understanding of how biology works is not necessary if you are to use that biology merely as a tool. While this may seem at first glance questionable, it is how we operate with all technology. I type these words onto the keyboard of a computer whose hardware is largely mysterious to me, using software written in a language as foreign to me as Mandarin. Even the simplest mechanical tool, such as a monkey wrench, is made by a process that requires expertise unknown to most people who use it. With the application of

5. The University of Heidelberg team picked up the 2013 Grand Prize. They designed a system that assembles short strings of amino acids for potential use in a range of schemes, including the recovery of gold from disused computer parts. The official project name was "Foundational Advance in Peptide Synthesis," but it was better known as "the Philosopher's Stone."

engineering principles to biological components, the question is not how these parts work, but whether they work. This way, biology becomes nonexclusive. By standardizing the parts, the entry-level criteria are lowered past the point where traditional DNA expertise is required. Paul Freemont runs the Centre for Synthetic Biology and Innovation, Europe's largest synthetic biology department, at Imperial College London (which will host the biannual premier synthetic biology meeting in 2013 called SB6.0). He told the BBC in 2012 that they have "a whole load of new people coming in, young, energetic researchers. They don't care where the biology is, they just want to build things [and] they want to solve problems. They don't have any baggage."

Who Owns Your Creation?

Because the purpose of synthetic biology is to create solutions for real-world problems, the potential for commercialization is obvious. Inevitably, the question of ownership looms over any components, devices, and products that get made. The story of patents and DNA is a tortuous and ongoing one, messy not least of all because the laws regarding patenting were not designed with biotechnology in mind. The promise that synthetic biology holds for future technologies means that the question of patents is intrinsically entwined with the science, and may yet have an effect on how this necessarily creative industry will flourish. Therefore, it's worth having a mercifully brief look at the problems.

Patents are historically designed to protect useful functional inventions, whereas copyright protects the physical expression of an idea—be it in the form of a written or music manuscript, a photograph, or a web page—by granting ownership, usually to the creator.

But more fundamentally, both copyright law and patents were introduced in the United States in order to foster creativity for the greater good. Inventors need protection over their creations, as the time, money, and intellectual investment they put into designing and making a new product is considerably greater than is required for simply copying it.

So-called products of nature, on the other hand, are ineligible for patents, as they occur without any human ingenuity. Yet it has been successfully argued that once a genetic sequence, such as a gene, is isolated and characterized, it is no longer a product of nature, as it has been copied and subjected to the intervention of humans. Therefore, it can be protected by a patent. The most high-profile recent test case of DNA ownership was (and continues to be) for two breast and ovarian cancer genes, BRCA1 and BRCA2, both of which were owned by the company Myriad Genetics and the University of Utah Research Foundation. On June 13, 2013, in a hugely significant decision, the U.S. Supreme Court issued a decree that Myriad's ownership of these genes was invalid on the grounds that simply isolating something that was found in nature did not qualify them as being patentable. The back-and-forth about DNA patents rumbles on. Nevertheless—primarily using the argument that the isolation of genetic sequences is a process—currently one in five human genes, the genes that you carry around in your cells, are effectively owned by someone else, for the time being at least.

When it comes to patenting living things, the law, at least in the United States, is marginally clearer. The first patent on an organism was upheld in 1980, with a genetically modified bacteria that helped break down crude oil. It was granted, on appeal, on the grounds that its genes were at least partially man-made, and that the process of manufacture was patentable. The key phrases used in the ruling were that the invention (and therefore the patent) constitutes a "manufacture" or "composition of matter." In 1988, this was extended to a multicelluar organism: the OncoMouse is a transgenic mouse whose DNA has been modified to make it particularly susceptible to various cancers, and therefore is a valuable research tool. And in 2010, Craig Venter filed a patent for the genome of his artificial cell Synthia.

Because the products of synthetic biology have little historical precedent, it is not yet clear how the law will (or will evolve to) handle them. In this case the complexities of DNA patenting highlighted above are compounded by the fact that much of the thought and, indeed, behavior behind synthetic biology components resembles, often by design, computer software. It gets even murkier here, as software is a field also beset by continual messy ownership debacles. Software falls somewhere between patent and copyright, as it is both functional (it enacts a program) and an intangible work (it is written to be enacted). The behavior and use of individual parts or pieces of software source code, such as formulas and calculating machines, can change depending on how they are used. As a result, computing patents are sometimes bafflingly broad, encompassing very nonspecific terms.

My wading into this dark world of intellectual property has specific relevance to the nature of the BioBricks part. As a result of the nebulous legality of biotechnology, broad patents have been used to grant intellectual property rights over mechanisms of basic cellular machinery. They cover general principles as well as specific processes. This has the effect of restricting experiment on those cellular activities, especially for researchers at the beginning of their careers, who might not be able to afford to pay for patented processes. If broad patents could be applied to the devices that synthetic biologists have created, such as over an oscillator in a generic sense, then it could potentially limit modification of what has become a basic instrument in the synthetic biology toolbox.

The BioBricks Foundation has recognized these potential pitfalls and taken a very precise stance on them. They state that its mission is "to ensure that the engineering of biology is conducted in an open and ethical manner to benefit all people and the planet. We believe fundamental scientific knowledge belongs to all of us and must be freely available for ethical, open innovation." They have reflected that patenting and copyright law as applied to computing was riddled with inadequacies. The BioBricks Foundation user agreement has enshrined an irrevocable protection from the assertion of intellectual property rights. Attribution to the creator and adherence to safety practices and, naturally, the law, are part of the agreement, but the part itself is quintessentially free to use. By placing their wares in the public domain, the users are precluding the possibility of thickets of patent enforcement, and consequently encouraging innovation and creativity.

In April 2012, recognizing the potential that synthetic biology has to change the world, President Barack Obama launched the National Bioeconomy Blueprint, a roadmap of how to invest in synthetic biology in order to extract the maximum benefit from it. This plan has commercial development written into it, but also recognizes the value of the BioBricks Foundation's core beliefs about sharing data and resources. Very specifically, this set of recommendations is designed to promote growth in synthetic biology, address safety issues, encourage commercialization without restricting creativity, and ensure the public is coming along for the ride.

The protection of creativity is enshrined in the titles of patent and copyright legislation, both from 1790: the Copyright Act is subtitled "An Act for the encouragement of learning" and the Patent Act's mandate is "to promote [. . .] useful Arts." In music, computing, genetics, and potentially the newcomer synthetic biology, these principles have been mired in the wake of speedy technological advances that far outpace changes in the law. As a result, in biotechnology, patents are in danger of becoming the embodiment of the impediment of creativity. When that happens, progress halts.

Again, the parallels with music are striking. Sampling in recorded pop music has become virtually ubiquitous. Hip-hop, which is based inherently on sampling, grew to become the biggest and most dominant business in music. As corporations recognized its immense commercial potential, the companies that owned copyrighted music also began to exercise and enforce control over the sampling on which hip-hop and other music genres were based. Copyright laws were refined and introduced; the upshot is that unless you have deep pockets, nowadays it is virtually impossible to make music with the same creative freedom that helped define this genre.

For synthetic biology, wrangling with this tangled legal thicket is all part of the rapid maturation of a very young field. In 2011, one of its godfathers, Drew Endy, opened a new venture to provide support and an open legal framework for the continued development of biological devices, circuits, and tools. At the launch he declared that "[w]e now need to move beyond Lego metaphors and genetic toys to professional technologies." The message is that synthetic biology needs to mature, both scientifically and legally, and to become the global industry its founders think it can be. Yet it should also continue to be fueled by unbridled creativity and even playfulness, as this is the fertile ground from which the most innovative ideas grow. In the words of the Harvard law professor and copyright activist Larry Lessig, "A culture free to borrow and build on the past is culturally richer than a controlled one."

The creativity built into the BioBricks Foundation, and synthetic biology in general, fosters a different ethos from the field of genetic modification that preceded it. It also has an engineering principle at its heart. Both of these aspects build up the case for seeing synthetic biology as an industrial revolution at its inception. Yet science occurs as part of culture, not distinct from it, and, as we have seen with the emerging issues of ownership, these new biotechnologies face not just scientific or practical problems, but the challenges of bringing them into society.

CHAPTER 11

The Case for Progress

"It is easier to fixate on the threat than embrace the opportunity."

Rob Carlson, 2011

The threat of violence was not fulfilled. In fact, what was billed as being an event marked by vandalism and direct action turned out to be, by all accounts, much like a pleasant family day in the English countryside. On May 27, 2012, following a few weeks of publicity, negotiation, and public pleas, the clash between anti-genetically-modified-foods activists and scientists happened on a sunny Sunday in the green fields of Rothamsted, Hertfordshire, featuring ice cream, singing, and some attempts at rousing rhetoric. The activists Take the Flour Back had declared in their propaganda that there isn't enough scientific data to support farming GM (genetically modified) crops, and that trials of GM wheat in Rothamsted's fields should be halted and the crops destroyed. They had declared that on this day they would "decontaminate" the fields of GM wheat, as has happened on several occasions in the last few years in the United Kingdom, meaning that they would rip out the experimental crops and raze the fields. The scientists leading the trials publicly pleaded and debated with them, but the plan of direct action stayed in place.

The wheat in the field carries several genetic modifications. Some are for lab administrative purposes (so that the functional experimental seeds can be selected from others). The main man-made gadget, however, is the introduction of a genetic circuit that produces the chemical (E)-β farnesene, or EβF (which, coincidentally, is very closely related to the chemicals being made for synthetic diesel mentioned in chapter 8). EβF acts as a Klaxon for aphids, a pheromone they pump out when being assaulted by predators such as ants; it tells other aphids to head for the hills. Aphids produce EβF themselves, but in a classy evolutionary move more than four hundred plants that aphids like to occupy also produce it, as a sort of BEWARE OF THE DOG sign on a canine-free house. Not only does it repel the aphids, it also prompts them to give birth to winged offspring, which can flee the perceived threat of danger. The experimental wheat has been created to produce EβF with the specific purpose of reducing the amount of aphid-slaughtering pesticide that farmers normally need to use.

One earlier study had shown that EβF-engineered wheat had little or no effect on repelling aphids, but that was under lab conditions, which can be qualitatively different from plants in an open field. The Rothamsted trial was designed as an attempt to address that exact question. A plaintive plea in an open letter by the scientists conducting the research pointed out that what Take the Flour Back had planned was "reminiscent of clearing books from a library because you wish to stop other people finding out what they contain."

Despite the decidedly nondramatic unfolding of events in Rothamsted that day, Take the Flour Back remained unrepentant and issued a statement afterward saying,

We wanted to do the responsible thing and remove the threat of GM contamination; sadly it wasn't possible to do that effectively today. However, we stand arm in arm with farmers and growers from around the world, who are prepared to risk their freedom to stop the imposition of GM crops.

This small saga reflects the state of play in the public perception of modern biotechnology, and in many ways its entire forty-year history. Take the Flour Back is vocal and noisy, and generates plenty of publicity. But it is difficult to gauge to what extent this group of activists genuinely echoes the public mood or the numbers of the farmers with whom they claim solidarity. In many ways they draw inspiration from the Luddites. Though the name has evolved to refer to anyone who rejects technology, the first Luddites were artisan cloth makers who felt threatened by the advent of new mechanical looms. In the space of a few years at the beginning of the nineteenth century, they met by moonlight on the Yorkshire moors in northern England, broke into factories, and smashed up the looms that had taken their jobs.[1]

For opponents of genetic modification, this kind of direct action is rare, but not unheard of: Take the Flour Back's website claims that "between 1999 and 2003 at least 91 GM trials were damaged or destroyed." The sentiment behind this has roots as old as the technology itself, as we shall see below. In the United Kingdom the potential introduction of genetically modified food during the 1980s was met with horror from campaigners, some of whom felt that to mess with nature in such a way was morally and intrinsically wrong. The press

---

1. That particular movement was short-lived but weighty, and was crushed with ruthless force. Legislation was introduced in 1812 that specifically outlawed vandalism to machinery and made it a capital crime. Seventeen Luddites were hanged a year later, and many were sent to what was then the penal colony of Australia.

fueled the revulsion by creating the term *frankenfood*—a peculiar portmanteau moniker, given that Dr. Frankenstein was the creator of Mary Shelley's unnamed monster, not the monster itself. Yet labels like that have power. Just as the 1972 thefts at the Watergate building in Washington DC inexplicably gave the press the green light to forever affix the suffix -*gate* onto any political scandal, in the last few years we have had headlines with frankenmice, frankenfish, frankenbugs, and frankencrops in reference to anything genetically modified.

As synthetic biology grows scientifically, it is drawing attention from all quarters and inheriting the antagonists of generic genetic engineering. In order to address the concerns and ire that synthetic biology raises, we therefore need to look at the much broader picture of the public perception of biotechnology in general, and its short but world-changing history. We are beginning to see groups that have historically led the charge against GM, such as Action Group on Erosion, Technology and Concentration (ETC) and Friends of the Earth, taking very specific aim at synthetic biology and publishing pamphlets alleging that the products of synthetic biology are untested, poorly understood, and that they threaten all manner of human and ecological scenarios. Immediately following the publication of Craig Venter's Synthia (it was in fact ETC that bestowed *Mycoplasma mycoides JCVI-syn1.0* with that more friendly nickname), President Obama commissioned a report from synthetic biology experts to assess any potential threats, challenges, and benefits from the whole field. The report, published in late 2010, addressed many of the issues by consulting with various publics and professional scientists, and drew up five categories by which to observe its progress: public beneficence, responsible stewardship, intellectual freedom and responsibility, democratic deliberation, and justice and fairness.

In assessing with these criteria, the commission decided that existing regulation was sufficient to allow synthetic biology to grow safely for the maximum benefit, while minimizing any potential threat. The principle of "prudent vigilance" is cited, which roughly translates as "let's keep a close eye on this." However, it also recommended openness and innovation through sharing, recognizing the open-access, patent-free ethos of the BioBricks project described in the previous chapter, and a coordinated approach across the whole field. It was a stimulus to a whole field that acknowledged the potential of synthetic biology and strove to nurture its growth.

Yet not everyone saw it like that. The commission drew considerable ire from campaigners. In response, fifty-six organizations led by ETC and Friends of the Earth immediately published an open letter to the presidential commission decrying the report for

> ignoring the precautionary principle, lacking adequate review of environmental risks, placing unwarranted faith in "suicide genes" and other technologies that provide no guarantee against the escape of synthetic organisms into the environment, and relying on industry "self regulation," which is the equivalent of no independent oversight.

Regulation for genetic modification is needed and does exist. In the United States three federal agencies rule over the use of GM plants: the United States Department of Agriculture presides over the possibility of crops becoming weeds; the Food and Drug Administration is the authority on their entering and subverting the food chain; and the Environmental Protection Agency judges plants with pesticide powers.

In 2012, the GM opponents increased their numbers to a coalition of 111 activist groups and put together their own report, which further denounced the field of synthetic biology. That number may sound like a lot, but many of these organizations are small groups of a handful of individuals, so it is difficult to assess how representative their view is of the general population. Again with charged language and robust assertions, their conclusions were that more oversight and more regulation were needed, and that there should be a full moratorium on the release and commercial use of synthetic cells.

It might be useful to look here to the origin of biotechnology to help understand the context in which this opposition is rooted. The epicenter of synthetic biology is at Stanford University, where the BioBricks Foundation is headquartered, and in the mid-1970s it was also the birthplace of the founding technology. The discovery of restriction enzymes, those bacterial cut-and-paste tools with which this whole story began, had enabled an experiment by Paul Berg where he cut out parts of the genome of one virus and inserted them into another. This baby step into the world of genetic engineering earned him a Nobel Prize in 1980. Yet he was cautious about the implications of this new technology; he held back from completing the experiment for fear of creating a monster and putting his colleagues and the wider public at risk. Instead, Berg called for a moratorium on this infant field. In 1975, at the behest of the U.S. National Academy of Sciences, he and others drew together a small international conference in Asilomar on California's Monterey Peninsula.

The Asilomar meeting is now regarded as a paragon of scientific responsibility when dealing with new and dual-use technologies, that is, new tools that have the potential to be used for both positive applications and by malefactors. The background culture in which it occurred featured the paranoid terror of blockbuster films such as *The Andromeda Strain*, in which an alien microbe caused madness and death indiscriminately, tapping into society's post-Watergate anxiety. Berg and others who were beginning to tinker with genetic code met with a sense of open caution in discussing where to go next. Journalists, lawyers, and scientists also convened, and together they hashed out the arguments over several days. In the end they drew up guidelines that suggested new research, open dialogues, and drawing on expertise from experts beyond molecular biology, such as in infectious diseases and microbial ecology. They concluded that the "new techniques, which permit combination of genetic information from very different organisms, place us in an area of biology with many unknowns." That is a sentiment that applies as much today as it did then. In fact, the nature of science in general should foment by definition a culture filled with unknowns. But the sense of potential harm was judged to be outweighed by the potential benefits, and they recommended qualified progress:

> It was agreed that there are certain experiments in which the potential risks are of such a serious nature that they ought not to be done with presently available containment facilities. In the longer term, serious problems may arise in the large-scale application of this methodology in industry, medicine, and agriculture. But it was also recognized that future research and experience may show that many of the potential biohazards are less serious and/or less probable than we now suspect.

However, the ideological rift between biotechnologists and their opponents was formalized shortly afterward. Just as in 2010, in 1978, not long after Asilomar, public opposition to the way genetic engineering was progressing began to emerge. There was also a falling-out of various scientists with environmental activists, including Friends of the Earth, with whom they had previously been allies. The terms were much the same as they are today: the degree of regulation required to enable safe progress in DNA-splicing research.

As the battle began, an anonymous scientist expressed the sentiment that "some of the environmental lobbies are in business to peddle paranoia," and that their behavior was merely thuggish. "I fear," James Watson told *Science* in 1978, "that such groups thrive on bad news, and the more the public worries about the environment, the more likely we are to keep providing them with the funds that they need to keep their organizations growing." Another scientist suggested, this time off the record, that "these private agents of the public interest are not elected, nor are they necessarily in touch with the views of rank-and-file members of the groups they speak for." It is quite possible that the same situation exists today, with popular support being garnered by shock tactics, publicity stunts, and emotive assertions rather than an attempt to engage in an honest and public dialogue, as we have seen in Rothamsted, and will see later in this chapter.

Paul Ehrlich, a prominent scientist and until that point a member of Friends of the Earth, reiterated the view that "the potential benefits from recombinant DNA research are so great that it would be foolhardy to restrict such research largely on the basis of imagined risks." It is true that no successful acts of bioterrorism have occurred since Asilomar, and we will come to that shortly, but as the industry of genetic engineering bloomed and mutated into synthetic biology, the potential problems remained the same, and access to the technology is considerably easier now than it was in the mid-1970s. In 1975 you could fit every single expert on recombinant DNA into a decent-size pub, let alone a conference center. The same technology has now been democratized to the point where everyone who has ever performed DNA manipulation would populate a small country. Romanticizing the past is probably not very helpful, but it seems that the climate in which the birth of molecular biology took place was significantly different. Biotechnology was a neonate then, not the scientific and commercial behemoth that it is today. A U.S. presidential report in April 2012 put the revenue of the nation's "bioeconomy" in 2010 at $100 billion. It is only set to grow.

The current actions and motivations of Take the Flour Back, and the policies of Friends of the Earth and ETC, are not significantly different from those expressed forty years ago at the birth of genetic engineering. But the field has exploded. The question, then, is whether the original opponents were prescient in their opposition, or if their rhetoric is as irrelevant as it is unchanged, when the science has progressed. We'll now look at some of the major claims and scenarios.

Biological Weapons

The specific case of weaponizing engineered life-forms is one of the key arguments put forward in opposition to both genetic engineering and synthetic biology. Our advancing skills in DNA technology have kept the fear of bioterrorism very much alive, as well. Weaponizing life is not a new phenomenon. The use of living things (other than humans) to create havoc has a rich but utterly ignoble tradition that predates our conquest of DNA by millennia. Hannibal used his attack elephants in Roman antiquity. A Mongolian chieftain invading India in the fourteenth century dispatched camels ablaze to charge scimitar-wielding elephants. Even in the modern era, rats, cats, pigeons, and dogs have all been used as either bomb-detecting or bomb-bearing devices. In the Second World War, the Russians had some arguable success training dogs to run under German panzers while carrying explosives triggered by a knock to the tanks' underbelly.

The potential threat of living weapons is no less bizarre nowadays, but far more potent, and, as with all aspects of genetic engineering, the key difference is the introduction of precision control of the language of life. Much like most biology, bioterrorism has been reduced in size from the use of animals to the microcosm of genetics.

The universality of genetic code is something that viruses have exploited throughout life's history, as they can insert their own DNA into an unwitting genome to usurp that host cell's own biological mechanics. We live in a world that has always been plagued by disease. Death by infectious agents far outnumbers death at the hands of another human, and our mastery of genetic manipulation has presented the possibility of terrorism with these tools. In 2006, reporters from the *Guardian* attempted to make a point by acquiring parts of the smallpox genome. This is a disease that has been eradicated from Earth.[2] Its genome sequence is freely available online, as all publicly funded genome sequences are, and the intention was to question how easy it would be to build a maleficent creation. By ordering a few sections of the genome of *Variola major,* the virus that causes smallpox, it was suggested that a whole genome could be stitched together from smaller pieces. The short, seventy-five–base sections were manufactured by an unremarkable UK gene-synthesis company and sent to a residential address in North London, unchecked for their weapon potential. As responsible investigative journalists, they introduced deliberate errors, and the sequences ordered encoded part of the virus that was itself not toxic. Nevertheless, the story ran with startling revelations:

> DNA sequences from some of the most deadly pathogens known to man can be bought over the internet, the *Guardian* has discovered. In an investigation which shows the ease with which terrorist organizations could obtain the basic ingredients of biological weapons, this newspaper obtained a short sequence of smallpox DNA.

---

2. Two smallpox samples are kept in cold storage for scientific purposes; there is an ongoing discussion about whether these should be kept or destroyed, primarily because of the risk of bioterrorism that they embody.

Shocking though this might sound, it was a ham-fisted stunt at best. Its only real triumph was to show that some DNA synthesis companies were not particularly vigilant in taking orders from members of the public. Also, their claim to have acted responsibly by building errors into the code was lip service. Once you'd gotten hold of the sequence, correcting the errors would be a trivial matter for any competent genetics graduate student.

Yet the overall premise was shaky. Using this technique of stitching together 75-letter fragments to make up a 186,000-letter genome is a tall order, akin to reassembling a shredded document twice the size of this book. "Theoretically possible," the article asserts, but in reality paralyzingly arduous. In 2002, a New York–based team did successfully assemble a working polio virus from mail-order synthesized segments of DNA. However, that genome is only 7,500 bases long and still took a team of molecular biology experts two years. These are not directly scalable problems. Recall that Venter's Synthia in 2010, with its 582,000-letter genome, took twenty people ten years at an estimated cost of $40 million. There is a wide range of actuality when discussing the theoretically possible.

Anti-synthetic biology activists ETC produced a report on synthetic biology in 2007 that refers to the *Guardian*'s stunt, which managed to enhance the theoretical possibility of further danger: "A commercial outfit could theoretically crank out the entire DNA for a synthetic version of *Variola major* in less than two weeks, for about the price of a high-end sports car." This might well be theoretically possible, but only if its meaning is stretched a long way beyond the practically feasible. As synthetic biology expert Rob Carlson says at the beginning of this chapter, "It is easier to fixate on the threat than embrace the opportunity."

It is important to acknowledge that there is a potential threat, and that synthetic biology and genetic engineering are dual-use technologies. Defunct diseases *can* now be reconstructed in the lab, and ones that we haven't conquered yet can be engineered to be more dangerous. We are capable of eradicating diseases such as smallpox, now consigned to history thanks to vaccination. Polio is likely to be next, and in time, diseases that kill us today will only be of historical interest to our children. Though it is unequivocal that science and medicine have transformed humanity's survival, it is equally apparent that millions die each year from incurable diseases. Yet how realistic is the possibility of genetic engineering's being the tool that malefactors use to bring the terror? *The Guardian*'s smallpox story ran six years ago, and as we have seen, the technology in question is evolving at a breathtaking rate. In 2012, the game changed yet again.

Killer Flu

Influenza comes primarily from birds. The virus annexes a host cell's workings to enact its own genetic program and make its own proteins. In flu viruses, two of these proteins sit on their surface as they drift from victim to victim: hemagglutinin (H), which hooks onto a target cell to gain entry, and neuraminidase (N), which newly made virus particles use to get out again. Variations in this pair of break-in and bust-out proteins give different flu strains their names, such H5N4. For the most part, H5N4 flu remains a bird disease. That's not to say it doesn't cause symptoms, just that humans won't make new viruses, so we can't spread it. But it is constantly evolving and finding new ways to spread. Once in a while, a strain will make the evolutionary leap to become a human disease, with both symptoms and the ability to transmit from person to person, with sometimes apocalyptic results. In 1918, the flu pandemic, caused by an H1N1 strain, infected billions and caused the deaths of fifty million people.

In 1997, patients in Hong Kong started dying from a new strain that hadn't been found in humans before: H5N1. This was a version that abounded in Chinese open-air chicken markets, to the extent that it was considered endemic. The very real worry was that, as it was a new strain, humans wouldn't have adapted any immunity to it. Flu's accelerated evolution can happen when two different strains infect the same cell, where they can trade genes. Therefore, if H5N1 swapped genes with a strain that struck humans and was transmissible from person to person, we would be looking at the beginning of disaster.

Knowing full well that highly infectious chimera strains are a theoretical possibility, teams of flu researchers set about fulfilling Feynman's maxim "What I cannot create, I do not understand." Scientists in the Netherlands and North America set about preempting nature by attempting to assemble the new combinations exactly as they might evolve to brandish pandemic-causing qualities. Yoshihiro Kawaoka from the University of Wisconsin–Madison and Ron Fouchier from Erasmus Medical Center in Rotterdam each designed flu viruses that would infect mammals, in this case ferrets. Kawaoka's design shuffled two decks of viral genes to put together one that not only would break into cells in a ferret's nasal passages, but could replicate and infect other ferrets via airborne droplets. Fouchier's design was similar but had just five genetic alterations, all seen in other natural strains, which enabled the virus to clamp onto ferret cells and infect. These also were passed on to other ferrets via a sneeze. Neither strain was lethal, though Fouchier's creation caused death when delivered in concentration via an inhaler. Given the importance of this work, these two studies were sent to the two most important scientific journals in the world: Kawaoka's to *Nature,* and Fouchier's to *Science*. What followed was a tortuous tale of negotiating a delicate path through frontier territory.

The U.S. National Science Advisory Board for Biosecurity

(NSABB) has oversight on exactly these types of scenario. NSABB is a collection of scientists, lawyers, and policymakers who advise on dual-use research in biology. Although it has no power to compel, in December 2011, NSABB recommended that the two papers should be published but redacted to "not include the methodological and other details that could enable replication of the experiments by those who would seek to do harm." This initial judgment prompted some serious and necessary hand-wringing by all parties involved. When controversial science is reported, the debates mostly tend to be between researchers and nonscientists. But this time around, unusually, there was intemperate disagreement between scientists on whether to publish in full, redacted, or not at all. As had happened at Asilomar, this confrontation prompted a dialogue open to the public. Also emulating the actions of Paul Berg in 1975, Fouchier and Kawaoka both called for a sixty-day voluntary moratorium on flu research in order to work out the right thing to do. The World Health Organization (WHO) stepped in, and after much gnashing of teeth the NSABB changed its mind, recommending full publication of the Kawaoka paper and a clarifying edit of Fouchier's. By this time, Nature had reached the same decision independently.

These studies show the very real possibilities of new and dangerous flu viruses emerging in the real world. They also show that to weaponize a flu virus would be a feat of genetic engineering, not something that anyone could do with any degree of ease; those studies took major investment and tens of thousands of highly skilled working hours. For a potential terrorist, this would not be the only disadvantage: a transmissible mutant mammalian flu virus could not be targeted at any one group of people, so any attempt to murder a specific population would be impossible.

Nor do viruses respect geography or nationalities. One theory behind the global spread of the devastating 1918 flu was that it began at a poultry farm in Kansas that supplied a local military base with chickens, and that the soldiers then took it with them as they were sent off to fight the first global war. These days, with world travel and a virus emerging in a dense conurbation such as Hong Kong, the potential for reaching humans all over the world is even greater. There is a threat from accidental release of transmissible viruses from a lab as well as malevolent intent. Even if you were hell-bent on creating a synthetic indiscriminate killer virus, rather than the millions it would take to make, the very well-proven process of artificial selection would be much more efficient. You could evolve the virus from chickens in a farm with dubious hygiene. Poor poultry-farming practices are far more likely breeding grounds for killer flus.

The utility of flu as a weapon is also questionable. If mass murder is your goal, then flying a passenger plane into a skyscraper is considerably easier. But the whole point of synthetic biology is to reduce the entry level for artisan genetic engineering, and the open culture of the BioBricks project has meant that gene manipulation, impossible a decade ago, is now school science. Does this make the entry level for bioterrorism lower? Currently, there are no bricks in the library that could be directly associated with a biological weapon. Yet it is possible to build a nail bomb out of components that are individually innocuous enough and can be bought from a standard hardware store. Indeed, one can make all kinds of explosives using simple household items, if one were so inclined. ETC's 2007 report on synthetic biology describes it as "genetic engineering on steroids," which ultimately will mean "cheaper and widely accessible tools to build bioweapons, virulent pathogens and artificial organisms that could pose grave threats to people and the planet." Maybe there is a flicker of truth in that sentiment, in that any progress in synthetic biology will allow the development of new technology with dual-use potential. Currently, though, and indeed for the foreseeable future, the reality of a genetically engineered weapon relies on the most generous use of the words *theoretically possible*.

This tale of the killer-flu experiments also raises a very fundamental question about the nature of science. One of the key principles in scientific research is the freedom to research any subject and for the results to be published. Perhaps with the advent of such potent dual-use technologies, this era is drawing to a close. But to censor scientific information about potential agents of terror is to ignore two problems. The first is that by studying how pathogens work, we understand better how to deal with them. This might be in the face of a terrorist threat, or simply to help treat patients, especially when virulence harbors epidemic potential. This study of pathogens needs to happen in an open, unrestricted way, as the richest, most productive, and most creative science comes when information is unfettered by barriers and shared freely. In doing just that we equip ourselves to tackle exactly the same threat should it come at all. Knowledge of how new virulent or even weaponized life-forms work will ready us for both natural and created threats. On publication of the Kawaoka paper, an editorial in *Nature* ruled that "A paper that omits key results or methods disables subsequent research and peer review." The final point is also a practical one. Keeping the information under wraps is simply not enforceable. *Nature* was aware that there were many versions of the paper circulating outside the official channels of the publication process, which has become increasingly common in the Internet era. The editorial went on to say that it could not "imagine any mechanism or criterion by which to sensibly judge who should or should not be allowed to see the work."

In a sense, these two studies were the first and most obvious experiments on a flu threat. Evolution is tenacious, and life-forms strive for immortality. In terms of understanding how that will unfold, we are always playing catch-up. We cannot predict every possible natural threat but we should do what we can to arm ourselves against the ones we can anticipate. Any knowledge in this arms race is better than none. That, to my mind, is good science, and good policy. It is using the precautionary principle in an aggressive way. It was unclear until these two flu studies whether it was even possible for these strains to jump the species barrier. They also provide clear genetic flags for public health sentries to look out for in emerging flu strains. By figuring out how to upgrade a flu into one that can cause a pandemic, we arm ourselves against the very real possibility of it happening naturally.

Agenda Politics

Another line of attack is that genetically modified food might be bad for us to eat. In biology we look for mechanisms that explain observations, and it's not easy to imagine what mechanism would result in an adverse effect on our health after eating a modified food. Nevertheless, if it were an observable phenomenon, it would warrant explanation. Opponents of genetic modification often cite the idea that GM foods might be bad for us, even though it has been thoroughly refuted in many independent reports. With many modified foods having been approved a long time ago, there is also more evidence than a scientist could wish for in demonstrating that they are safe to eat. According to one estimate, two trillion meals containing genetically modified food have been consumed in 2009 and 2010. The U.S. Department of Agriculture calculated that GM revenues in 2010 were $76 billion, with the products being part of the human food chain.

Despite this, the arguments continue. In September 2012, a small study published in a minor journal hit the headlines and stirred the

GM debate yet again. A French team of scientists, led by a molecular biologist named Gilles-Eric Séralini, fed a type of genetically modified maize to rats over their whole lifetimes and observed any effects on their health. The modified feed had been developed by biotech company Monsanto to resist a widely used herbicide. It has been widely approved as food for humans and animals around the world, and previous studies have shown no adverse health effects with this particular modified maize (called NK603, previously Roundup). Indeed, in 2012, a systematic review of multiple long-term studies, which is considered to be the gold standard in these types of trials, also concluded that there were no adverse health effects for a diet of GM food. In Séralini's study and the surrounding press, they claimed some of these previous studies were qualitatively different from theirs. Their paper graphically showed that the rats used were profoundly affected by their GM diet, suffering grotesque tumors and dying significantly before the control rats.

Immediately, though, the study began to unravel. The authors and the paper were bluntly criticized for the quality of the work, and also for the way it was reported. With a swiftness enabled by the Internet, scientists began condemning the data, saying that the experimental design, the statistical analysis, and the presentation of the data were all substandard. Some expressed surprise that such a paper was published at all. Specifically, critics noted that Séralini had used rats that were already prone to developing tumors, as well as inadequate numbers of control rats compared to experimental ones.

It soon emerged that the release of the results was part of an orchestrated PR campaign that included a book by Séralini, who has a track record of anti-biotech activism, and a television documentary. The paper itself was not freely distributed to the press before publication, which is extremely unusual and makes science reporters highly suspicious: most science journalists will seek comment from other scientists, albeit under embargo. With Séralini's paper, a signed agreement was required for those who were offered access to the paper, which came with another highly unusual stern warning: "A refund of the cost of the study of several million euros would be considered damages if the premature disclosure questioned the release of the study."

An organization called the Sustainable Food Trust (SFT) had spearheaded a campaign to spread the work as far and as wide as possible, including a dedicated web page resource, wanting to promote what it calls "Good Science." The SFT encouraged supporters to pass on prefabricated messages of support on the influential social media site Twitter. Almost all of the mainstream press had covered the paper, although interestingly, some of the coverage, such as by the BBC, had included some of the immediate profound criticisms from scientists. A few days later, the European Food Safety Authority (EFSA) issued a statement: "The design, reporting and analysis of the study, as outlined in the paper, are inadequate," adding in the press release that the paper is "of insufficient scientific quality to be considered as valid for risk assessment."

But the PR offensive had done its work. Many press outlets published the story uncritically. The French weekly newsmagazine *Le Nouvel Observateur* shouted out on its cover, *"Oui, les OGM sont des poisons!"* ("Yes, GMOs are poisonous!"). The French prime minister declared that France would defend banning these genetically modified crops across Europe (though he had the good grace to specifically add the caveat of needing to validate the results). Bearing in mind the distinctly unusual way the paper was published, the results are not invalid until other studies show them to be. That is not to say that they are correct; given the strong criticism of the methodology itself, it seems fair to say that at the very least, Séralini's study demands the most rigorous replication of the results, releasing the original data for independent analysis, and shoring up the methodology to establish beyond doubt its validity.[3]

Within minutes of publication, many were forensically investigating not just the paper, but the way it was published. To get traction for an idea is the aim of such a PR exercise, and in this case they partially succeeded, in the face of the rapid scrutiny of critics. Some have suggested this case study marks a turning point in the conflict over GM, as the condemnation was so quick and so severe that some of the mainstream press included it. Just like rumors, once scientific ideas like this are out in the open, they are very difficult to correct, regardless of poor methodology, or knowing the agenda of the perpetrators, or anything else.

There are many volumes of publications on GM; in these pages I am mentioning the tiniest handful from the very recent past. This is not to cherry-pick the data to fit a particular stance, a crime of agenda that I denounce, but to illustrate how research into GM is subject to manipulation and is thrust into the limelight because it evokes passionate and polarized responses. Public and political opinion can, and has been, skewed by the way these types of campaigns are reported; the deliberate polarization of public debates detracts from nuanced, informed, and necessary analysis of the application of biotechnology.

---

3.   To get published in a scientific journal requires peer review, in which other expert scientists rate the validity of the proposed publication and recommend it (or not) for publication. Yet it is not a hallmark of scientific truth, only a recognition that a group of scientists believe an experiment is worthy of entering the academic literature. In this case, some have disputed whether Séralini's paper should have been published at all. Historically, critical feedback came via the cumbersome method of publishing correspondence or replicating experiments in academic journals. Now there is a form of post-publication peer review enabled by the Internet, in which papers are scrutinized in blogs and columns and discussed openly. It's not formal and is subject to the wildness of a frontier without gatekeepers, but it is quick, and scrutiny was precise and harsh. By December 2012, though, the journal of the EFSA published a paper that concluded Séralini's study "was found to be inadequately designed, analyzed and reported," and "of insufficient scientific quality for safety assessments."

Malaria

To focus on the possibility of harm from GM foods or bioterrorism is to fixate on a threat, whether looming or currently implausible. As an industry instead poised to provide opportunities, the nascent field of synthetic biology has so far yielded few genuine success stories. In this book I have expressed enthusiasm for its potential, but in this chapter and elsewhere I have looked at some of the harder realities of applying groundbreaking science into society, which is explicit in synthetic biology's mandate. There have been innumerable examples of the scientific successes of genetic engineering in its broadest sense, not least in terms of understanding the basics of biology and the underlying causes of thousands of diseases. The work I once did in the lab helped identify a form of blindness in children, and this would not have been possible without the genetic modification of mice.

Synthetic biology has yet to deliver on that scale. However, it is set to start, and just like any discipline that involves extremely complex and constantly changing science, it is subject to the scrutiny of not just people and politics, but also market forces and economics. The story of the synthetic biology company Amyris and their stalled attempt to turn synthetic biodiesel to market is described on pages 156–59. Probably the single biggest success story of synthetic biology, however, along with the quagmire of real markets, comes from exactly the same team, labs, and genetic circuit board.

Throughout human history, malaria has been a constant lethal presence. Some estimates put the number of people who have perished at the hands of plasmodium, the malaria parasite ferried by mosquitoes, in the tens of billions, depending on how you define human species. Currently, a quarter of a billion people become infected every year, and the highest estimates are that more than a million will die, mostly young children. Apart from the staggering human cost, the WHO estimates that the cost of malaria to gross domestic product for sub-Saharan Africa since the 1960s has reached $100 billion dollars. The impetus to reduce the burden of malaria is profound.

Since the seventeenth century, quinine, extracted from South American cinchona trees, was the preferred treatment, though it came with a suite of unpleasant side effects. A chemical cousin called chloroquine usurped quinine as the standard antimalarial drug following the Second World War. However, as with all life-forms, the parasite wants to continue to live, and it does so by evolving. Mass treatment with chloroquine resulted in the 1950s in the emergence of plasmodium strains that were resistant to chloroquine, and spread across the world in a bid to ensure their own survival in the face of man-made extinction.

The current drug of choice in malarial areas is a small molecule called artemisinin. It is extracted from sweet wormwood, an Asian camphorous shrub, which is grown all around the world and has been used in folk medicine for centuries. As a treatment for malaria, artemisinin is fast and effective. The WHO specifically prohibits its use on its own, for fear of unintentionally nurturing and selecting strains of plasmodium that are naturally resistant. Instead, the organization recommends that artemisinin be the key player in combined therapies.

Getting enough artemisinin is a pricey and finicky business. It has to be farmed quickly and in very specific ways. From an economic point of view, the very fact that it is cultivated means that the fields of wormwood are competing with crops for food or feedstock.

Drugs are created in chemistry labs. You start with basic chemical ingredients, which can be bought from chemical companies or harvested from the living world. The production of drugs is like

forensically accurate cooking, as each ingredient is mixed in to systematically add or subtract bits of molecules until you have the drug
desired. Artemisinin is not a large molecule, but unfortunately it's
also not easy—or more important, cheap—to synthesize using
straightforward chemistry. Early in the process of working on synthetic circuits to create biosynthesized diesel, Stanford University
researcher Jay Keasling was alerted by a student to the fact that one
of the steps in their chemical pathway was also a key link in the
chain that could produce synthetic artemisinin. While he tried to
build a genetic circuit that would create diesel, his team also pur-
sued another that would construct artemisinin. They published the
first successful circuit for artemisinin synthesis in yeast in 2006,
having switched from bacteria in 2003. It's built from twelve genes
taken from three different organisms.

In both cases upscaling to industrial production was inherent in
Keasling's aims, again marking the intent not merely to create, but
to directly apply scientific research to real-world scenarios. It's a
strange thing to see in a scientific paper, but from the start reducing
the cost of production was clearly set out as a mission: "Industrial
scale-up will be required to raise artemisinic acid production to a
level high enough to reduce artemisinin combination therapies to
significantly below their current prices." This exemplifies the
problem-solving, engineering sensibility built into synthetic biology. It denotes the application as the goal, not merely to produce a
functional drug, but also to do it inexpensively.

Keasling's rationale for this apparent altruism stems from the
market forces that had confused artemisinin farming in the preced-
ing years. At the end of the twentieth century, there was a great
paucity of wormwood farming, so the market drove the price of
artemisinin up. Then, spotting a gap in the market, thousands of
African and Asian farmers started growing wormwood. The price
dropped. Yet even with the drug being artificially maintained by

government clinics at $1 a dose, more than half of patients would buy more expensive versions out of convenience from market stalls not acknowledging the government intervention. Unreliable supply, fluctuations in prices, and variable yields drove the prices back up, and this pattern looked set to continue.

Keasling set up Amyris to develop artemisinin production, along with its diesel sibling (which has tanked, at least temporarily). Following that, the Bill and Melinda Gates Foundation, the charity set up by the Microsoft billionaire and his wife, chipped in with $46 million to the charity Institute for OneWorld Health, which worked with Amyris to realize industrial-scale production. Amyris has run this project on a not-for-profit basis and joined up with organizations such as the Global Fund to reduce the cost and distribute as much artemisinin (as part of a combination therapy) as possible, as cheaply as possible. That means driving the cost down to less than fifty cents per treatment. Amyris granted the pharmaceutical giant Sanofi-Aventis a royalty-free license to produce vats of synthetic artemisinin, which are expected to hit the shelves within the next two years.

Implementation of a cocktail of antimalarial drugs is the WHO's strategy for combating malaria. This way, resistance doesn't evolve against artemisinin and declaw its therapeutic power, as it has for its ancestors. Yet companies disregard the WHO's strategy. Because the constituents of combined therapies can be sold individually, there is profit to be made in ignoring this organization's advice. In 2009, the WHO felt compelled to issue a briefing that warned companies to stop marketing artemisinin as a single therapy. According to the WHO report, the motivation of these companies is strictly financial. Andrea Bosman, part of the WHO's Global Malaria Programme, told *Nature*: "It's terrible. Who says there is no profit to be made in malaria? When you see the number of companies operating in Africa, and the diversity of products, you'd just be amazed."

The reality of applied science in the marketplace provides the major caveat to this story. Exactly what WHO scientists were trying to avoid came to pass. The first cases of artemisinin resistance had been discovered in Myanmar in 2004, following in the footsteps of chloroquine resistance in the mid-twentieth century. Containment of the evolved parasite became the priority, but by 2012, patients with limited responses to artemisinin were recorded in Cambodia, indicating that the resistance had spread (or possibly evolved separately). With luck, the principles of component switching and circuit tweaking that characterize synthetic biology will make the road to modifying malaria treatment easier and smarter in the years to come.

In the fledgling artemisinin farming market, prices have fluctuated wildly in cycles of boom and bust. Though this instability is undesirable, part of the argument against the establishment of a synthetic production base is that it would result in the displacement of thousands of wormwood farmers, as their livelihood would be ruined by a greatly reduced cost of synthetic production. This is one of the attacks that anti-synthetic biology campaigning groups put forward, and it may have some legitimacy. In a 2012 briefing pamphlet specifically criticizing the synthetic artemisinin project, the ETC reported that artemisinin demand could be met solely with an increase in wormwood cultivation, and therefore synthetic production was not only unnecessary but would also take production control from small-holding farmers and hand it to large Western pharmaceutical corporations.

Part of this argument runs the risk of falling foul of what's known as the Luddite fallacy, the idea that technological unemployment follows technological advances, as first mechanization and then automation replace workers. It's a fallacy, at least in the long term, because if it were true, as economist Alex Tabarrok points out, "we would all be out of work because productivity has been increasing for two centuries." At the same time, the combination of a not-for-profit enterprise to produce an affordable drug to treat millions of people must outweigh those concerns. The collaboration between Amyris, OneWorld Health, and Sanofi-Aventis bears the hallmarks of corporate responsibility. They have agreed to work on a not-for-profit basis and to limit supply to 50 percent of the full artemisinin market, which leaves some space for traditional farming. Whether this collaboration will be the benchmark of humanitarian aid enabled by synthetic biology is yet to be seen, because at the time of this writing, synthetic artemisinin has not hit the markets. If and when it does, artemisinin will be the first genuine commercial product of synthetic biology and, if successful, it will be puzzled over, studied, and followed for years.

Should or Could? The Case for Progress

Science is primarily a public endeavor, not least because it is largely funded by the public purse, but also because the benefits of non-commercial scientific research are for all. As synthetic biology emerges, it will face all of the concerns that GM has had over the short history of biotechnology, and more as it grows. One thing seems to be essential: the discussions about synthetic biology and genetic modification must happen in public, and with the public.

The calls for bans from the opponents of genetic modification and now synthetic biology are unrealistic and destructive. They are designed to foment fear derived from an ideological position, to enrage rather than engage. The eighteenth-century satirist Jonathan Swift suggested it was not possible to reason someone out of a position that they didn't reason themselves into. If resistance to new forms of biotechnology are ideological, then contesting those views with evidence is a daunting challenge. There are legitimate and serious concerns in any of the technologies described in these pages, but they demand rational, open, and informed discussions.

GM crops are in the wild and they are in our food. Another significant challenge from green campaigners is that, once out in fields, GM crops can cross-breed with traditionally manipulated crops and the unnatural genes can outcompete their more natural counterparts in the open marketplace of nature. This is a more legitimate concern, as we know that this can happen. As with the EβF wheat in Rothamsted, gene engineering in crops is frequently aimed at reducing the need for pesticides by getting the plant to produce its own. Similarly, engineering a crop to be resistant to a weed killer is potentially useful, as it means that farmers can spray their fields with an herbicide knowing that only unwanted plants will die. But so far, field experiments have produced mixed results. Some GM crops do appear to have the effect of reducing local biodiversity, with fewer wild plants and fewer insects to pollinate them. However, it is not universal. In one experiment GM maize appeared to increase biodiversity, whereas beets and rape did the opposite. That sort of result is not uncommon: biology is messy. But this should lead us to conduct *more* experiments, not to shut down or even vandalize them.

More scientists and politicians than ever are now moving toward the stance that, in order to address mammoth population growth, poverty, and incoming climate change, genetically modified crops will be utterly necessary. Sir John Beddington, the UK government's chief scientific adviser (GCSA), told the BBC in 2011 that

> If there are genetically modified organisms that actually solve problems that we can't solve in other ways, and are shown to be safe from a human health point of view, and safe from an environmental point of view, and they can solve problems we can't solve otherwise, then we should use them.

The precautionary principle is carefully threaded into that statement, suggesting the necessity of GM rather than a whimsical or commercial desire to introduce these creations. There is a simple necessity underlying Beddingfield's comment: in changing environments, many of which are going to make land in poor nations less amenable to being cultivated, we need new ways for crops to be hardier in difficult ground. Breeding is slow and cumbersome, and therefore is not a realistic option in our timeframe.

For synthetic biology, its potential for use in the service of humankind and the planet puts it in a similar camp. We don't know the full effects of introducing its products out of the lab and into the wild. One of the main demands of synthetic biology antagonists is that these cells and unnatural life-forms should not be released. We have seen with GM crops the spread of their modified genes beyond the original host, though whether these have a detrimental effect is unknown. Could any of the products of synthetic biology cause havoc? The cancer assassin circuit described in chapter 9 integrates into the genome of a virus and delivers its lethal message after infecting a cancerous cell. It is designed to target only one specific type of malignant cell and performs a calculation to precisely determine its cancerous nature. If this circuit were capable of going native and joining the natural ecosystem, would there be a risk of its entering other cells and destroying them? It is unimaginable that it could, because of the very precise nature of the design of the circuit, yet it is theoretically not impossible, just tremendously improbable. We do know that genes swap in and out from microorganisms and viruses, and this can't be ruled out.

Could Synthia live beyond the lab? *Mycoplasma mycoides* is a natural pathogen, albeit a minor, nonlethal one for goats. It might be able to survive outside the lab, though Venter's team specifically wrote into its genome a chunk of code that rendered it incapable of infection, thus denying the pathogen its favored habitat. Yet the remarkable tenacity of bacteria suggests that it could gain functions upon exposure to other bacteria, as they swap their genes.

Could Amyris's artemisinin- or biofuel-producing yeast cells escape from their vats and wreak havoc on ecosystems? It is unlikely, as they, too, are optimized for purpose, not for survival. They are not in the wild, or in field tests, though their nonliving products will be.

All of these scenarios are not airtight, but after careful consideration the risks seem almost trivial, and just like with the outstanding questions in genetically modified farming, they suggest the need for more research, not less.

Knowledge Is Value Free

In 2007, an editorial in the journal *Nature* made this comment: "Many a technology has at some time or another been deemed an affront to God, but perhaps none invites the accusation as directly as synthetic biology. For the first time, God has competition." It was written in broad support of synthetic biology, but carries a theological misunderstanding. Ever since Eve took a bite of an apple, God, if one believes in such things, has had competition. Throughout our existence we have challenged, manipulated, and forged nature to our own ends. Our influence has shaped and defined this living world, hitherto governed by Darwinian rules. With synthetic biology we have the opportunity not to supplant those rules, but to create new lives specifically for purpose. That is not a call to defeat nature, nor to trample it underfoot any more than we have already done. In less than a fraction of a heartbeat of this planet's existence, we have done more than enough to jeopardize our continued existence here. The living planet will go on cycling according to the fixed law of gravity, with or without us, Earth's most creative and destructive offspring. Yet with smart, informed scientific decisions we have the ability to fix our past errors.

In 2005, on the thirtieth anniversary of the Asilomar meeting, Paul Berg commented, "First and foremost, we gained the public's trust, for it was the very scientists who were most involved in the work and had every incentive to be left free to pursue their dream that called attention to the risks inherent in the experiments they were doing." One-tenth of the conference attendees were journalists who were free to observe the sometimes bitter altercations and fights between scientists. It was the scientists who raised the concerns and, by arguing about them out in public, they ensured that the outcomes were cautious, progressive, and uncontroversial.

It may be that current regulations will not be sufficient in the future, but they should be addressed at that point, not preemptively in a way that could prohibit progress. If prudent vigilance is built into the structure of research funding, public engagement, and the way the research is done and applied, then the precautionary principle is foremost in the way synthetic biology proceeds. Whether this works remains to be seen, but to halt progress is to deny the potential benefit that new technologies can provide. Furthermore, increased regulation, restricted progress, and heavy monitoring drives actual research out of the hands of publicly funded scientists, because all of these things are expensive. Those levels of bureaucracy are handled comfortably by companies with commercial interests, and they are not obligated to have open dialogues with a public who funds them.

Synthetic biology is moving at such a fast pace that many scientists are bewildered by its progress. To bring the public and politicians on board is a difficult task. Yet it is essential, as it is society that makes the decision about what we should and shouldn't do. "Democratic deliberation" is the phrase used in President Obama's 2010 bioeconomy report. To this end, in the United Kingdom in 2010, major scientific funding agencies commissioned extensive research on what people know and think about synthetic biology. Key questions emerged:

> What is the purpose?
> Why do you want to do it?
> What are you going to gain from it?
> What else is it going to do?
> How do you know you are right?

These are precisely the right questions any scientist should ask of any project, so in that respect this type of dialogue is working.

Yet in reflecting on the triumph of the Asilomar meeting, Paul Berg also warned that an attempt to repeat past successes would be a futile gesture: "By contrast, the issues that challenge us today are qualitatively different. They are often beset with economic self-interest and increasingly by nearly irreconcilable ethical and religious conflicts and challenges to deeply held social values. An Asilomar-type conference trying to contend with such contentious views is, I believe, doomed to acrimony and policy stagnation, neither of which advances the cause of finding a solution."

I don't share this pessimism. I believe that scientific research should occur in the full glare of public scrutiny and that scientists should engage with publics of all levels of expertise. This way, with data out in the open, and informed public conversations about the potential benefit and the potential harm that new technologies create, we foster a society in which rational approaches to global and local problems are normalized.

There is a revolution happening and it is in our hands. Synthetic biology promises benefits for all humankind's residency on Earth and on worlds not yet explored, far too great to ignore, repress, or censor. There is youthful optimism, a remix culture of boundless creativity that believes these new technologies will help rectify the problems we face, and at least in some quarters, there is an unprecedented willingness to push that agenda.

We will face new and unforeseen challenges, some of which we have created or at least nurtured. Yet we should strive to invent technologies that are not in conflict with nature, nor subvert or exploit it, but work alongside our sophisticated living world with its four-billion-year-evolved history. Humankind's progress is born out of our attempts to explore and understand a continuously changing landscape, and live within those means. Our exploration of the workings of living things over two or three centuries has given us the power to do things that could never have happened before. We are engineering life-forms to build fuels, drugs, treatments, tools to explore our universe, and boundless new living creations that can help our world and our tenure on it. Our responsibility is not to curtail that knowledge but to use it to better ourselves and our living planet. Sometimes this gets framed as a question of whether we should do the things that we are capable of. The answer transcends either option: we must.

Afterword

New Languages

anguage evolves. Samuel Johnson noted in his famous dictionary that all languages "have a natural tendency to degeneration," though throughout history terminal decline has been somehow miraculously avoided. That anxiety about the way languages change—along with a suggestion that languages once, at some point in the past, existed in perfect and inviolate form—has been expressed in every generation throughout history by brabbling malagrugs, but really only reflects the fact that words and their meanings constantly change over time. Cromulent words are added, mangled, misappropriated, and invented. Very occasionally, languages acquire new letters, too.[1]

Evolution settled on an alphabet of just four letters and twenty

---

1. Almost always, this pilfering is to cope with words on loan from foreign languages. French has virtually no uses for the letter *w* except in borrowed words, such as *"le weekend."* Welsh acquired *j* to cope with English and Norman words and names. Languages and their alphabets contract and expand like living things.

words, and just a single language. This alphabet and lexicon[2] has remained static since the beginning of life on Earth, a stable mechanism that is flexible enough to encourage evolution in a species but shields the individual from catastrophic changes to their genomes.

With the genetic code being the only language of life we know of, there is no possibility of importing letters or words from other life-forms; no direct analogy for the kind of borrowing and exchange we see between languages. However, genes and DNA—the words in our analogy, as opposed to the letters of the genetic code—can swap between species, particularly singled-celled species of bacteria and archaea. Yet they can also be transported harmlessly into more complex creatures when viruses integrate their genomes into a host. It is estimated that up to 8 percent of your DNA was once in the genome of a virus.

Just as we have difficulty in pinpointing the precise moment when a species comes into being, we cannot pinpoint exactly when English was born, or any of the myriad languages or dialects that are and have been spoken throughout human history. However, *Kelkaj lingvoj ne evoluas; kelkaj simple elpensiĝas*, as Esperanto speakers might say: some languages did not evolve; they were simply invented. Esperanto was a noble attempt to induce world peace by making a universal language; it survives a century after its creation with a few thousand speakers. In fact, something like nine hundred languages have been invented from scratch, including Klingon, as spoken by the angry fictional warriors of the *Star Trek* franchise and a few hundred devoted followers.

Our command of the language of DNA and its mechanics has brought us to the point where we no longer need be content with the four-billion-year-old alphabet, lexicon, and language of life. In the

---

2. *A, G, C,* and *T* (*U* in RNA) arrange in genes into triplets, which together encode one of just twenty amino acids and three stop codons.

era of synthetic biology, we have begun the process of inventing new ones.

If you've ever had a cold sore, you've probably already used invented DNA letters. The universal letters, nucleobases, are merely chemicals. A for adenine is a collection of fifteen atoms: five carbon atoms, five nitrogens, and five hydrogens. They're arranged into a hexagon attached to a pentagon with a few molecular bobbles protruding. There is nothing inherently special about that arrangement, nor of the other bases that make up the genetic code. That doesn't mean that they are easy for us to build into that vital configuration, as John Sutherland's work demonstrated in part I (pp. 106–8), but those limitations are ours and not nature's. At some point in the distant past these chemicals acquired meaning—began to arrange themselves into a system from which information could be derived, stored, and handed down. The system arises from the shape of the molecules; how they link with one another and form a chain. We can now make our own chemicals, similar enough to these naturally occurring ones that they can be integrated into DNA. We're already using them in the fight against virus infections.

Typically, a virus will storm into a cell carrying its own genetic code but none of the mechanics to translate it. One of the reasons viruses are not typically classed as living is because they can only reproduce by using someone else's kit. So viruses hijack a living cell, insert their own code, and hope that it doesn't notice. This steganography by the virus uses the same language, so cells often don't spot the inveigled plan and they read it unwittingly. As a result, the host cell produces more viruses, which break out to infect more cells, often destroying the host in the process. Antiviral treatments include false letters that are similar enough to be incorporated next to their natural counterparts but conspicuous enough to disrupt any meaningful message. If you spot a massive interloping typo in the middle of a sent&ence, your brain is sophisticated enough to skip over it without misunderstanding. The cell is not so forgiving of the alien letters. They are designed so that the cell's translation mechanics see the new bases as unnatural, and won't skip over them. This is how many antiviral and cancer chemotherapy drugs work. These man-made alien bases are used as standard treatments for infections from HIV to herpes: when you dab cream onto your lips to soothe a cold sore you are probably applying acyclovir. The balm contains a molecule similar enough to the base *G* to go into the replicating virus code, but different enough to stop the sentence from continuing. The usefulness of this Scrabble bag of alien letters comes from the very fact that they are alien. They don't act as a readable code. Instead, their insensibility to cellular machinery gives them power to rudely thwart the virus's sole desire—to copy itself. Reproduction is halted, making it a very useful tool for fighting these invaders.

Naturally, the most interesting things we can do with invented DNA rely on our creating code that *can* be copied. If a code can't be copied, then it can't be passed on, and if it can't be translated, it can't have a function (beyond the terminating one described above). Many attempts have been made to make letters that can be snuck into DNA, and not only stay there, but get copied. Synthetic biologist Steve Benner (Westheimer Institute of Science and Technology) has led this effort and, in 2011, added two alien bases to the alphabet of genetics.

The double helix of DNA has two spines that twist into a spiral, so it has two grooves on the outside—twin spiraling slides. One is wider than the other, so they are called the major and minor grooves. Electrons whiz around the atoms down in the minor groove, just as they do in all molecules. Here they form an electronic pattern that acts like a chemical identifier, and they play an essential role in DNA replication. The protein that copies genetic code, DNA polymerase, has evolved to recognize this particular electronic signature, and is drawn in to begin its copying work. In order to trick the cell into incorporating unnatural bases into DNA, Steve Benner's tactic was to design new ones, called Z and P, that mimic the electron signature when integrated into the double helix, and DNA polymerase gets into the groove. We now have a DNA whose code is made up of an alphabet consisting of A, T, C, G, Z, and P. DNA polymerase happily does its natural thing, reading the code and copying it. As always, the copying process is not perfect, and there is an error rate equivalent to the accumulation of the mutations in DNA that drive evolution. Benner has added brand-new letters to the genetic code.

These new letters don't mean anything yet. This design and construction is a proof of principle—that extra letters can be added to DNA and that they will not disrupt the cell's normal behavior. The new techniques of biological engineering tend to rely on the existing mechanics of the cell to translate and enact the new designs of synthetic biologists. Naturally, alphabet letters don't have any inherent meaning in isolation, either. It's only when strung together into agreed-upon words and phrases that they stop being mere noise and transform into prose that can express love and render the complex understandable. If we were to simply add new letters to the English alphabet, say Љ and Ц (from Cyrillic), we would have no way of pronouncing them, and they would not have any meaning until a consensus was arrived at. However, this is not the only way to invent new languages.

Alien Code

This focus on the letters of the spiraling ladder of DNA is only one aspect of synthetic biology's attempts to reengineer the alphabet of life. DNA is a complex molecule, a polymer, meaning that it is assembled from repeating units. The code is hidden in the rungs of the ladder, but another major breakthrough in alternative, unnatural genetics has come not from swapping out or inventing new rungs, but the uprights of the ladder. These are made from a type of sugar called deoxyribose—the D in DNA (in RNA it is simply ribose)—which repeat and link up to form the spines of each strand. Since 2000, biologists have been creating alternative spines from a range of other sugars to produce several new genetic molecules: ANA (with arabinose), TNA (with threose), and four others including FANA and CeNA. These alien species are collectively known as xeno-nucleic acids, or XNAs. The letters of code remain the same, but the struts of the ladder are strange. In April 2012, again at the Laboratory of Molecular Biology in Cambridge, a team led by Philipp Holliger and Vitor Pinheiro built a system that for the first time enabled these foreign-language genes not only to replicate, but to evolve.

The genius of this experiment was not how the code was put together but the fact that it gets copied. DNA polymerase, the protein whose job is to copy a single strand of DNA to make it into a double, does so by reading one strand and picking up and stringing together the pieces to make that strand's mirror, A for T, C for G, and so on. DNA polymerase won't attach to anything other than DNA, understandably, as it has evolved over billions of years to do only that, and in nature there is nothing for it to act on but DNA. Proteins such as DNA polymerase are long and complicated, and their function is determined by their three-dimensional shape (which is determined by the order of their amino acids, which is determined by their genetic code). Because of this intricate 3-D conformation, they resist intelligent design: it is immensely difficult to redesign something so complex. Evolution is blind but clever; it designs by random experimentation and by selection of successful mutants. Pinheiro presided over a process of natural selection by creating a pool of polymerase proteins that were all fractionally mutated in key areas. By selecting mutants that could pick up XNA pieces instead of DNA, he forcefully evolved a polymerase that reads DNA, but lays down a mirror strand using XNA, as if the alien molecule was native. In this way, the brand-new, human-invented genetic molecule can carry the same message encoded in natural DNA. They effectively tricked the machinery of cellular cryptography into reading its natural language but copying it into an unnatural one. They built a tool that will translate a natural language into a made-up one, English into Klingon.

Yet that is only half the story. They can also translate back again. Pinheiro mutated and selected a protein that would do the opposite, translate from XNA back to DNA. That is a trickier trick, as no natural protein will do the job. After eight rounds of forced random mutation and selection, they had crafted something that did, which makes XNA a system bearing two hallmarks of life: information storage and heredity. Fidelity is important when copying, especially in genetics, but perfect fidelity is not evolution, it is stasis. Their process is faithful at around the 95 percent mark, which means the stored information can evolve.

This, if one cares about such naming conventions, marks the inception of a new branch of biology—synthetic genetics. Admittedly, this first study still relies on DNA as its template, but the team is already trying to cut it out of the loop. They have copied FANA to FANA and CeNA to CeNA, though it doesn't work quite as well as DNA yet.

The implications are weighty. Pinheiro and his colleagues have shown that genetic evolution is not limited to the natural code as we know it. There may well be uses for this gallimaufry of XNAs in the future, as they can behave like natural genetic parts even though they are not, and might not be treated as such by our immune systems, which have several billion years' worth of schooling in recognizing natural DNA- or RNA-based genetics. XNAs are more robust than DNA and RNA, both of which are prone to breakages and cuts by proteins called nucleases, whose role is to do just that. By being immune to slicing, XNAs have extraordinary therapeutic potential. Geneticist and origin-of-life scientist Jack Szostak has developed a class of molecule called an aptamer, which is strands of DNA or RNA designed to fold in such a way as to clasp onto a very specific target. As a potential therapy, this action has the ability to shut down a gene or inactivate a mutant protein. Currently, there is only one aptamer on the market, as a treatment for a degenerative eye disease. Yet it is vulnerable to being chopped up by nucleases patrolling for rogue bits of DNA, so the medicine has to be taken repeatedly. Theoretical at present, an equivalent XNA aptamer would be invisible to nuclease, a stealth weapon in the armory of medicine.

Altered Output

Both of these projects reinvent the code of genetics. Naturally, there is the output of that code to consider as well. Life is built by or of proteins, and with the advent of sophisticated molecular biology, proteins are now also subject to rewriting.

Proteins are built from the nose-to-tail joining of amino acids. At one end an amino acid will have an arrangement of atoms called an amine group (which is made up of a nitrogen atom and two hydrogens), and at the other a group that forms carbolic acid (which is a carbon atom, two oxygens, and a hydrogen). The genetic code in

DNA spells out twenty amino acids, which are collected together in the process of protein manufacture and assembled by the cell.[3]

Amino acid is a generic name for a set of molecules that are all quite similar. They vary only in what is attached to the middle section of the molecule, between the amine and the carbolic acid. The simplest amino acid is glycine, with only a single hydrogen atom as its unique side chain. At the other end of the scale is tryptophan, with a large double ring of carbons, hydrogens, and a nitrogen sprouting from the side. All types of variation in between make up life's lexicon. These side chains determine the behavior of the protein in which the amino acids are assembled. It's important to remember that there's no real limit to the number of amino acids that could exist, in theory, beyond the twenty that life actually uses.

To get a sense of how this system has been breached by scientists in the last few years, we first need to understand how the machinery of the cell works. Protein construction is remarkably like a mechanical production line. The gene being translated is first transcribed from DNA into a free-floating copy in RNA, maybe a thousand bases long, untethered to the rest of the DNA of the host cell. This RNA message drifts to one of many ribosomes, the center where the protein will be constructed. In the cellular milieu around the ribosome float the amino acids that will be strung together according to the order set out in the messenger (or m-) RNA, each one specified by a three-base codon—the specific triplet of letters that encodes an amino acid. Several key molecular players shift these molecules around in the manner of a factory production line, and the process

---

3. In fact, our bodies make just eleven of these amino acids, and we need to ingest a further nine. There are two other amino acids life-forms infrequently use, one of which is exclusive to bacteria and archaea. In general, we tend only to refer to the canonical twenty.

works as follows. The gene to be translated (in the form of mRNA) is fed into the middle of the ribosome, like feeding paper ribbon into a ticker-tape machine. The RNA bases are read codon by codon, and the corresponding amino acids are added one at a time. The protein is ejected like the ticker tape being spat out. This output is the basic protein, to be folded and transported to its site of use. Yet the agents in this process are effectively manufactory handlers. The ribosome itself is of fundamental importance: its shape and construction, largely out of RNA itself, is the forge into which all the ingredients trundle along this biological conveyor belt. There is also a complex of three molecules that very specifically deliver the amino acids to the ribosome. This combination is basic to all life, but not easy to explain. One part is the amino acid itself, which is collected by the second part, a folded piece of RNA that specifically collects and transfers it to the ribosome. For this reason, these are generically called transfer (or t-) RNA, and work by bearing the anticodon, that is, bases complementary to the codon itself—a *T* to pick up an *A,* a *C* to pick up a *G,* and so on (though there are only twenty amino acids, there are dozens if not hundreds of tRNAs). The final handler is a protein with the unwieldy name of aminoacyl tRNA synthetase, which I shall refer to as synthetase henceforth. This protein clamps the amino acid to its corresponding tRNA, and together the component parts of the protein are delivered for assembly. The code input goes in at one end, the ingredients are delivered from another as a package, and the protein rumbles out, one amino acid at a time.

This process occurs in all life-forms and its universality is one of the cornerstones of the evidence for a unique origin of life. The components are clearly ancient—no matter how distantly related two species are they share the basic workings of this assembly plant. To change it has required an example of evolutionary brute force, an intelligent design pioneered by Peter Schultz at Scripps in California, and by Jason Chin and his team at the LMB in Cambridge.

The existing code has plenty of redundancy built into it: there are sixty-four possible combinations of the four bases, and sixty-one of them are used to encode the twenty amino acids of life.[4] That leaves three remaining possible combinations, and in life-forms all of them bear the message STOP, that is, the instruction to terminate protein production. These punctuations are essential to indicate the end of a gene. Indeed, the significance of the stop codon is demonstrated in a handful of genetic diseases: Sandhoff disease, lethal in infancy, is caused by a gene bearing a premature stop codon, like a sentence cut incomprehensibly short by a rogue period.

One of them, the triplet *UAG*, is called the amber stop.[5] Jason Chin has appropriated this sequence as a codon for building amino acids otherwise unrecognizable to living cells. There is no tRNA-synthetase pairing that will bind to an amber stop, and so Chin's idea is to evolve one. Chin and his team start with this pairing from a species distantly related to the experimental cell. This ensures the pair won't start interacting with the cell's normal workings: the mechanics of protein production are universal, but that doesn't mean that the kit is exactly the same. In order to get the alien synthetase to singularly recognize something that none of the other tools of life can—the amber stop—they use the basic premise of evolution. If for some reason you wanted a fish that only ate something artificial such as jellybeans, you would start with a densely populated pool and use a jellybean as bait. If there was a fish in that pool that, unusually, had a natural taste for jellybeans, you would catch it with that bait. You could then breed a new pool from that fish and create

---

4.   For example, six different combinations encode leucine: *UUA, UUG, CUA, CUG, CUC,* and *CUT*; see part I, chapter 5 for the reasons for the evolution of this redundancy.

5.   Amber stop is not named in relation to the intermediate warning at traffic lights, but because its discoverer's name was Bernstein, German for "amber."

a population that liked jellybeans. By repeating the process itera-
tively, you would end up with fish that only fed on jellybeans.

The fishing exercise is similar to the process of evolving a tRNA-
synthetase pair that will recognize unnatural amino acids. By starting with a pool of versions of synthetase, which have randomly
mutated parts in the region that locks onto the amino acid, you can
select ones that bind to your new unnatural amino acid, and repeat
the process. Eventually, just like the jellybean-eating-fish, you have
created a tRNA-synthetase that will only read a stop and an unnatural amino acid.

That's one part of it. The tRNA is bred in that way to act as the
anticodon for the amber stop. Adding a *UAG* into the sequence of a
gene so it can be picked up by the tRNA is an awkward process, but
essentially uses the tools of DNA manipulation not unlike those
used for any gene tinkering—the creation of Freckles the spider-
goat, for instance. You can synthesize a whole gene sequence, rewritten from scratch, complete with an amber stop at a crucial
point. Or you can use a copying technique called mutagenic PCR
that will copy imperfectly and introduce the new triplet in error.
UAG normally only exists in a gene to indicate its end, so in order
for this system to work, all the elements have to be changed: the
code, the synthetase, and the tRNA, all to recognize an unnatural
amino acid.

This is delicate biological engineering to the most fundamental
living process—the "central dogma," as Francis Crick described it.
But it is not mere experimentation for the sake of it. Jason Chin and
others are using it to discover how proteins interact with one another. Living cells are a network of protein interactions: some attach
themselves to DNA, some to the molecules of metabolism, and frequently they connect with other proteins to perform their vital
functions. The unnatural amino acid incorporated into Chin's experiments carries a side chain that does two things. First, when

instructed to do so, it will form a locked bond with the protein it is interacting with, which means that we can fish the pair of them out and identify the unknown partner, with the engineered protein acting as bait. Second, the unnatural amino acid is light activated. To get it to lock onto the target, you merely shine ultraviolet light onto the cell, and the bond is sealed.

Chin is not the only person who can perform this act, but his team is the first to get it into multiple species, including animals. Until 2012, these unnatural actions were restricted to cells in culture, in dishes where the environment is controlled and neat. Chin has incorporated this process into fruit flies, one of the standard animals of experimental biology, and shown that no aspect of biology is off-limits when it comes to reengineering: alphabet, code, proteins—the whole system of life is now ripe for rewriting.

These rewriters show that while our system, the only one we know of, is powerful enough to form the basis of all life on Earth, it doesn't have to be that way. Evolution by natural selection was discovered and described a long time before we understood its mechanics. The code that enacts that process of error and trial is unique, but is quite capable of being adjusted, rewritten, or even invented afresh. It's difficult to conceive of a system of biological evolution (that doesn't rely on some form of supernatural creator) other than natural selection, but we only have one system of coded, reproducible information that has worked. XNA, the bases Z and P, and amino acids that are unnatural to the cell all show that there can be others. This is the first baby step toward the creation of new life-forms that use a Darwinian genetic code invented not by nature, but fully by us.

Digital DNA

With these examples we can see how the encroachment of synthetic biology on nature has given us the power to radically alter the genetic code of DNA, to invent new versions of genetic material, and even to expand the lexicon of genetics so that it can include unnatural proteins.

The language of DNA is not restricted to the creation of life. Our genomes are data storage devices, imperfect by natural design to encourage adaptation to the changing environment. Yet DNA is remarkably stable; it has transmitted information in the same form for billions of years, and the information held in DNA remains intact long after the death of the cell or organism that bears it. More and more frequently these days stories emerge from the new scientific world of ancient DNA. The genome of the wooly mammoth has been published—extracted and decoded from hairs sixty thousand years old, bought on eBay by the geneticists who sequenced it. Our likely evolutionary cousins the Neanderthals joined the genome club in 2010 when their complete DNA was read from 44,000-year-old bone specimens. Yet the current record stands with DNA extracted from frozen mud cores in Greenland, identified as being from the family of perennial plants saxifrages, with an age range of between a whopping 450,000 and 800,000 years old. True, being buried beneath six and a half thousand feet of ice is about as close to a laboratory freezer as nature can manage, but the fact that we can find long-dead tissue and still decode the messages written within it shows what a stable data storage device DNA can be.

This is one of the reasons that scientists in the 1990s began thinking about DNA as a means of storing not just the biological information required for a cell to function but also digital data. Although not the first to take advantage of DNA's data storage talents, Craig Venter's synthetic bacteria, aka Synthia, is the most famous living digital device to date. As mentioned in the introduction, within the computer- and machine-synthesized genome of the goat pathogen *Mycoplasma mycoides,* Venter hid messages in the DNA that would be cryptic even to the machinery of the cell. Computer coders often hide secret treasures or messages—Easter eggs—in their programs to give credit, or simply as fun puzzles for users to discover.

There were four parts to these DNA Easter eggs. The first was a code table. The cryptic messages were all conceived in English; a cipher had to be included so that the four letters of genetic code could spell out the twenty-six letters of the English alphabet plus punctuation. The twenty amino acids are sometimes referred to by a single letter, *A* to *W,* so a conceivable way of concealing an English language message in DNA would be to simply use the existing genetic code in the triplets that spell out amino acids, three bases for each Roman letter. Instead, Venter constructed a new cipher for the English alphabet, so this needed to be deciphered. The code was cracked within a few days of publication.

The second and third concealed messages comprised the names of the few dozen creators of the cell and an Internet address. The final Easter egg was a set of three apposite quotations. The first was "To live, to err, to fall, to triumph, to recreate life out of life" from the novel *A Portrait of the Artist as a Young Man* by James Joyce.[6] The second quotation came from J. Robert Oppenheimer, the

6. As mentioned on page 140, it was the subject of a copyright challenge by James Joyce's estate.

so-called father of the atomic bomb: "See things not as they are, but as they might be." The third was an accidental misquotation of that marvelous phrase from Richard Feynman referred to in chapter 7: "What I cannot build, I cannot understand."[7]

These blocks of DNA sequence were designed by hand and assembled in a computer. They had to be flanked by DNA tags that indicated they were not to be translated by the cell, because the code they are written in is entirely new and would have been nonsense. That sequence is not written in the triplet codons of the genetic code; it cannot and will not make a protein. Instead, Venter had designed a storage device with an encrypted code that was invented and new, simply using the alphabet of DNA as code, without reference to its history.

In 2012, Harvard Medical School's George Church drove the commoditization of DNA into a new era with the first publication of an entire book encrypted digitally in DNA (entitled *Regenesis*, it

---

7. Which, in case you are wondering, looks like this in DNA:
TTAACTAGCTAATTTCATTGCTGATCACTGTAGATATAGTGCATTCTATA-
AGTCGCTCCCACAGGCTAGTGCTGCGCACGTTTTTCAGTGATAT-
TATCCTAGTGCTACATAACATCATAGTGCGTGATAAACCTGATACAATAG-
GTGATATCATAGCAACTGAACTGACGTTGCATAGCTCAACTGTGAT-
CAGTGATATAGATTCTGATACTATAGCAACGTTGCGTGATATTTTCAC-
TACTGGCTTGACTGTAGTGCATATGATAGTACGTCTAACTAGCATAAC-
TAGTGATAGTTATATTTCTATAGCTGTACATATTGTAATGCTGATAAC-
TAGTGATATAATCCAACTAGATAGTCCTGAACTGATCCCTATGCTAAC-
TAGTGATAAACTAACTGATACATCGTTCCTGCTACGTGATAGCTTCACT-
GAGTTCCATACATCGTCGTGCTTAAACATCAGTGATAACAC-
TATAGAGTTCATAGATACTGCATTAACTAGTGATATGACTG-
CAAATAGCTTGACGTTTTGCAGTCTAAAACAACGTGATAATTCTGTAGT-
GCTAGATACTATAGATTTCCTGCTAAGTGATAAGTCTACTGATTTACTA-
ATGAATAGCTTGGTTTTGGCATACACTGTGCGCTGCACTGGT-
GATAGCTTTTCGTTGATGAATAATTTCCCTAGCACTGTGCGTGATATGC-
TAGATTCTGTAGATAGGCTAAATTCGTCTACGTTTGTAGGTGATAGTT-
TAGTTGCTGTAACTAATATTATCCCTGTGCCGTTGCTAAGCTGTGATAT-
CATAGTGCTGCTAGATATGATAAGCAAACTAATAGAGTC-
GAGGGGGAGTCTCATAGTGAATACTGATATTTTAGTGCTGCCGTT-
GAATAAGTTCCCTGAACATTGTGATACTGATATTTTAGTGCTGCCGTT-
GAATATCCTGCATTTAACTAGCTTGATAGTGCATTCGAG-
GAATACCCATACTACTGTTTTCATAGCTAATTATAGGCTAACATTGC-
CAATAGTGCGGCGCGCCTTAACTAGCTAA.

is appropriately enough about synthetic biology). At 53,000 words, eleven images, and one script of software code, it is three-quarters the size of the book you are holding, or around five megabits of information. The code is very simple binary: an *A* or *C* represents a 1, and a *T* or *G* represents a 0. Words were converted into a digital form and then the equivalent sequence of DNA was synthesized in a computer in ninety-six base lengths, each with information tags about the location of the segments and so on—what coders call metadata. Fifty-five thousand of these fragments make up the book, roughly one per word. In translating the whole book into DNA, there are only ten errors in five million bits of data. The whole thing is stored on what's called a DNA chip, which as a means of storing a book is not that different from a page and ink, just much smaller. These chips are small glass plates about the size of a book of matches, coated with chemicals that DNA will stick to. The DNA data fragments are literally sprayed by a tiny nozzle from an inkjet printer. Unlike the information stored in Craig Venter's synthetic bacteria, Church's method never goes near a life-form. The DNA is written in a computer, synthesized by a machine, printed with a printer, and decoded by a machine and software using the very same standard techniques employed by scientists recovering DNA from millennia-dead tissue.

Like many of the invented techniques in this chapter, this is a proof of principle. As a means of information storage, DNA's resilience is irrefutable. George Church and others have taken the simple fact that DNA can carry a code and have rewritten it to store non-biological data. The stability of DNA and the falling costs of the technology used give this technique potential for archiving data at a breathtaking density: the authors claim 5.5 petabits per cubic millimeter (i.e., one quadrillion units of information). This is a more concentrated from of data storage than a Blu-ray Disc, a flash drive, or even the hard drive on your computer. What has been achieved so far is only useful for archiving, as the technology to write and access the memory stored in DNA takes many days; electrical circuitry in a computer can save an equivalent amount of data in seconds. Yet it is not inconceivable that one day computers might store their memories not in silicon chips, but in DNA chips.

That creative engineering philosophy is key to synthetic biology: how can we redesign and use biological technology for purpose? All of the endeavors described above are nascent technologies, still very much in the experimental phase. Yet they all demonstrate that the limits of nature are being surpassed by our own inventions. We have always adapted nature for our benefit, and in the era of molecular biology we have done so by remixing it at a molecular level. Now, for the first time, we are engineering living systems created in ways that rewrite the very language evolution has provided.

Acknowledgments

Just like science, writing this book has been a collaborative process. I wrote the words, but *Creation* is a concatenation and remix assembled from the ideas and work of others. My thinking on the subject of the origin of life is heavily influenced by my friend Nick Lane at University College London. I embarked on *Creation* before reading his outstanding book *Life Ascending*. But, having done so now, and for our many lunches, I am profoundly indebted to him. In the first half, I've liberally borrowed from his thoughts, which he expresses with virtuoso clarity. I urge you to read it.

Many of the interviews and much of the research for this book were drawn from my involvement in BBC radio and television productions. In 2009, I presented a documentary series on BBC4 called *The Cell*, which during the course of three hours covered the story of biology from van Leuwenhoek to synthetic biology. This was made by a team of brilliant filmmakers including Jacqui Smith, Andrew Thompson, Alison Rooper, Nick Jordan, and Jill Fullerton Smith. Particularly, David Briggs did significant research on the seventeenth-, eighteenth-, and nineteenth-century microscopists, and deserves acres of credit for academically and meticulously collating multiple original sources. In 2010, we followed this up with a

BBC Radio 4 *Frontiers* program on the current status of origin-of-life biology. Roland Pease produced and immense gratitude is due to him as well as to Sasha Feachem and Deborah Cohen at the BBC Radio Science Unit. During the writing of *Creation*, I presented an episode of the long-running BBC flagship science program *Horizon* on synthetic biology. Matthew Dyas, Kelly Neaves, and Aidan Laverty skilfully guided that into the world. Some of the interviews quoted in *Creation* were conducted under the auspices of those programs.

Here are some others to whom I owe a debt of gratitude: Jane Sowden, my friend and PhD supervisor, from whom I learned how to be a good scientist; Steve Jones, for teaching me, inspiring me, and appearing as a contributor in almost all of my television and radio programs; Armand Leroi, for helping me with Aristotle by showing me chapters from his forthcoming book *Aristotle's Lagoon*; Matt Ridley, for providing scientific and biographical information on Francis Crick; Rob Carlson, author of the excellent book *Biology is Technology*, for his help on many aspects of synthetic biology; Alex Taylor and Vitor Pinheiro, for much-needed guidance on XNA and expanded genetic code; John Sutherland, for RNA synthesis; Zita Martins, for her expertise on astrobiology and Martian rocks; Emma Perry, for helping me with the school curriculum; Sinjoro Simon Varwell, dankon; those on Twitter who have at times responded to my pleas to find the right word, fungus, or reptile, specifically @Mr._muse, @sarcasmaniac, @janieface1, @writerJames; Kirby Ferguson, for copyright and remixing; Richard Kelwick, for the UK iGEM meet up; Philip Campbell, Kerri Smith, Charlotte Stoddart, Geoff Marsh, and Thea Cunningham, and all my friends at *Nature*; The Celeriacs, semper smallage.

I was nomadic in writing, though rarely ventured out of Hackney, London. The following kind people allowed me to use their various spaces: Jack and Lynsey Mathew—and Barney for keeping me company (Lynsey also for the word "jumentous"); John Sanders and Beth Gibbon; Ana-Paula Lloyd and Jonny Hassid; and everyone at my second home, Mouse and de Lotz—quite clearly the best café in London.

Several people read and commented on the manuscript during its lengthy gestation. However, any errors or mutations that have survived these rounds of selection are mine alone. I wish to thank Ed Yong, Jim Al-Khalili, Matt Ridley, Dara Ó Briain, Matthew Cobb, Brian Cox, Suzi Gage, David Adam, Nathaniel Rutherford, Mark Lynas, and David Watson for their advice, conversation, and edits; and particularly Kevin Fong, with whom I have shared the chronic condition of writing. I tested many of the ideas in this book on him.

I am tremendously grateful to my agents, Sophie Laurimore, PJ Mark, and Will Francis; without Will's continual support *Creation* could never have happened; my wife Georgia and my children, who have supported its genesis unflinchingly; my love for them is as deep as geological time. Most of all, I humbly thank my editors at Penguin, Brooke Carey, Katie Coe, Kary Perez, and Will Hammond, who wrestled with me and these words countless times to batter them into shape. I doubt many writers receive as much attention in sculpting their prose.

Scientists can be a quarrelsome lot. They compete for ideas and for funding and, just like in any human endeavor, are subject to dogma and whimsy and hard-felt beliefs. Good theories emerge from that maelstrom of open—but sometimes fractious—debate, and get tested and refined by experiment. Science does not settle on truth, but tends toward it. These particular areas of science are pioneering and important and, as such, they are subject to quarrel perhaps more than most. The ideas presented here will be refined or changed over time, or rejected. And frankly, like all good scientists should, I look forward to being corrected.

Annotated Bibliography

We luxuriate in a time where the basic mechanisms of life on Earth are largely understood, and many great books have been written that describe evolution and genetics. My recent favorites include *Life Ascending* by Nick Lane (Profile Books, 2010), *Why Evolution Is True* by Jerry A. Coyne (Oxford University Press, 2010), *Your Inner Fish* by Neil Shubin (Penguin Books Limited, 2009), and almost everything by Steve Jones, particularly *The Language of the Genes* (HarperCollins, 1993; despite its being written twenty years ago, ten years before the Human Genome Project was first published, it contains insight and stories as good as any contemporary book on the nature of DNA), and *Almost Like a Whale* (Black Swan, 2000), which successfully updates the *Origin of Species* with contemporary evolutionary biology. Of course, reading anything by Charles Darwin will make you smarter. His works are as important to our culture as anything by any writer.

A Note on Latin, Plurals, and Language

Bacteria have existed longer and in greater number than any other group of organisms. Though invisible to the naked eye their total biomass exceeds that of both animals and plants. All other life on Earth is dependent on bacteria, including our own, as we carry trillions of these cells on and in our bodies.

Bacteria are also cornerstones of the era of molecular biology revolution, providing both the tools of genetic manipulation (restriction enzymes) and the medium in which almost all genetic engineering takes place.

In this book I have used the word *bacteria* to mean both single and multiple cells, rather than bacterium for a single cell. This is on the principle that its adoption into the ever-evolving English language releases it through usage from the grammatical rules of its neo-Latinate origin.

Part I: The Origin of Life

Chapter 1: Begotten, Not Created

As with all science, the pathways to great discoveries are complex, tortuous, and winding, and almost always bereft of eureka moments. The road that led from Van Leeuwenhoek to cell theory is no different, with dozens of people making small incremental experimental and conceptual advances. I have outlined the narrative of just some of the key players. For a thorough and scholarly history of the early microscopists, Professor Sir Henry Harris's *The Birth of the Cell* (Yale University Press, 2000) is the definitive guide, from which I have sourced many sections in the story of cell theory.

Animalia (*The History of Animals*) by Aristotle, 350 BCE, translated by D'Arcy Wentworth Thompson in 1910, is thoroughly readable and well worth a look. It is available for free online at http://classics.mit.edu/Aristotle/history_anim.html.

Chapter 2: Into One

Plenty has been written about Gregor Mendel, as befits one of the great scientists, and here is the original paper where he outlines what would become the laws of inheritance: G. Mendel, "Versuche über Pflanzen-hybriden," *Verhandlungen des Naturforschenden Vereines* 4, Abh. Brünn (1866), pp. 3–47, or in translation: "Experiments in Plant Hybridization," *Journal of the Royal Horticultural Society* 26 (1901), pp. 1–32.

That there were 115 copies of that paper, none of which made it into Darwin's library, is documented here: R. C. Olby, "Mendels Vorläufer: Kölreuter, Wichura, und Gärtner," *Folia Mendeliana* 21 (1986), pp. 49–67.

One point of Mendelian interest lies in reanalysis of his data by Ronald Fisher, one of the founders of modern evolutionary biology (along with J. B. S. Haldane, whom I'm told he could not stand to be in the same room with, and Sewall Wright). Fisher concluded that the statistical significance of Mendel's results meant the experiments must have been "falsified so as to agree closely with Mendel's expectations." He wasn't alleging that the laws are incorrect, merely that the results were precise enough to suggest massaging. This has been repeated often since, but the following paper by the geneticists Daniel Hartl and Daniel Fairbanks concludes that this allegation can be put down "because on closer analysis it has proved to be unsupported by convincing evidence": Daniel L. Hartl and Daniel J. Fairbanks, "Mud Sticks: On the Alleged Falsification of Mendel's Data," *Genetics* 175:3 (March 2007), pp. 975–79.

The story of twentieth-century biology and, indeed, science, orbits around probably the most significant scientific advance of that century—Crick and Watson's publication of the double helix. For more on the story of DNA, there are plenty of books available. James Watson's own account, *The Double Helix*, is a gripping read (reissued by Simon & Schuster, 2012), though is very clearly a personal account by a fabulous storyteller, and to my mind bears the signature of mythmaking. His treatment of Rosalind Franklin in that book is horrid.

Franklin's role in the story of DNA is frequently debated, often with the question, should she have gotten the Nobel Prize alongside Crick and Watson? There is a simple and absolute answer to that, which is no: the rules are that Nobel Prizes will not be given posthumously, and Franklin was prematurely dead by the time DNA was recognized by the awarding committee in 1962. Brenda Maddox's biography *Rosalind Franklin: The Dark Lady of DNA* (HarperCollins, 2003) is thrilling, comprehensive, and honest, portraying Franklin in a believable and not always flattering light. It is a marvelous book.

Francis Crick, as befits his genius, appears in several different sections of this book, as he continued to explore the nature of life after his and Watson's grand discovery in 1953. There isn't a better description of his life and works than Matt Ridley's *Francis Crick: Discoverer of the Genetic Code* (HarperCollins, 2011).

The BBC dramatization of the tale of DNA is entitled *Life Story* (known in the United States as *The Race for the Double Helix* or *Double Helix*) (1987), and is thoroughly watchable. It features Hollywood's favorite otherworldly quirky scientist actor, Jeff Goldblum, in the role of otherworldly quirky scientist Jim Watson. As an aside, Goldblum has also played a neurosurgeon (*The Adventures of Buckaroo Banzai Across the 8th Dimension*, 1984), a particle physicist (*The Fly*, 1986), a mathematician (*Jurassic Park*, 1993), a computer scientist and environmentalist (*Independence Day*, 1996), a canine immunologist (*Cats & Dogs*, 2001), and a marine biologist (*The Life Aquatic with Steve Zissou*, 2004). Can any other actor boast such a litany of scientists?

"We wish to suggest a structure for the salt of deoxyribonucleic acid (D.N. A.). This structure has novel features which are of considerable biological interest." So begins probably the most famous and significant research paper of the twentieth century: J. D. Watson and F. H. C. Crick, "A Structure for Deoxyribose Nucleic Acid," *Nature* 171 (April 25, 1953), pp. 737–38. This is available at www.nature.com/nature/dna50/archive.html, along with several other landmark papers by titans of biology from this golden age, including O. T. Avery, C. M. MacLeod, and M. McCarty, "Studies on the Chemical Nature of the Substance Inducing Transformation of Pneumococcal Types," *Journal of Experimental Medicine* 79 (1944), pp. 137–59.

Language is the most useful and frequently used metaphor for understanding and explaining the workings of DNA. Language also shares many similar evolutionary characteristics, as explained in this review by biologist Mark Pagel: "Human Language as a Culturally Transmitted Replicator," *Nature Reviews Genetics* 10 (2009), pp. 405–15 (doi:10.1038/nrg2560). For a terrifically enjoyable nonscientific romp though the evolution of words and language, read Guy Deutcher's *The Unfolding of Language: An Evolutionary Tour of Mankind's Greatest Invention* (Arrow, 2006).

For Douglas Theobald's statistical analysis of Luca's existence, see his "A Formal Test of the Theory of Universal Common Ancestry," *Nature* 465 (May 13, 2010), pp. 219–22 (doi:10.1038/nature09014); and a rebuttal: Takahiro Yonezawa and Masami Hasegawa, "Was the Universal Common Ancestry Proved?" *Nature* 468, E9 (December 16, 2010), pp. 219–22 (doi:10.1038/nature09482).

This paper describes a model that explores the idea that a preponderance of the bases CG might indicate a proximity to hot vents: Laurence D. Hurst and Alexa R. Merchant, "High Guanine-cytosine Content Is Not an Adaptation to High Temperature: A Comparative Analysis Among Prokaryotes," *Proceedings of the Royal Society B* 268 (2001), pp. 493–7 (doi:10.1098/rspb.2000.1397).

On some of the earliest fossils of cells: David Wacey et al., "Microfossils of Sulfur-metabolizing Cells in 3.4-billion-year-old Rocks of Western Australia," *Nature Geoscience* 4 (2011), pp. 698–702 (doi:10.1038/ngeo1238).

There are many papers on bacteria and archaea swapping genes rather than inheriting them from their cellular parents. One of the most graphic is: V. Kunin et al., "The Net of Life: Reconstructing the Microbial Phylogenetic Network," *Genome Research* 15 (2005), pp. 954–59, which features a necessarily baffling diagram of the tangled bank of the base of the tree of life. *New Scientist*, a magazine with a proud history, decided to focus its attention on this precise subject on the bicentennial of Darwin's birth in 2009. The bold cover title was "Darwin Was Wrong," though the articles inside were more nuanced about the subject of lateral or horizontal gene transfer. They explained in an accompanying editorial:

> As we celebrate the 200th anniversary of Darwin's birth, we await a third revolution that will see biology changed and strengthened. None of this should give succor to creationists, whose blinkered universe is doubtless already buzzing with the news that *"New Scientist* has announced Darwin was wrong." Expect to find excerpts ripped out of context and presented as evidence that biologists are deserting the theory of evolution en masse. They are not.

On seven separate occasions I was presented with this cover (or someone quoting it) at public meetings as part of attacks from particular religious groups who choose to ignore evidence and assert creationist doctrine.

Chapter 3: Hell on Earth

See Euan Nisbet's classic textbook, *The Young Earth: An Introduction to Archaean Geology* (Allen and Unwin, 1987), on the first couple of billion years of our planet.

This study suggests that comets delivered water to the earth and the moon: James P. Greenwood et al., "Hydrogen Isotope Ratios in Lunar Rocks Indicate Delivery of Cometary Water to the Moon," *Nature Geoscience* 4 (2011), pp. 79–82 (doi:10.1038/ngeo1050).

An interesting model that suggests microbial life could have survived underwater during the intense barrage of the Late Heavy Bombardment can be found in Oleg Abramov and Stephen J. Mojzsis's, "Microbial Habitability of the Hadean Earth During the Late Heavy Bombardment," *Nature* 459 (May 21, 2009), pp. 419–22 (doi:10.1038/nature08015); Lynn J. Rothschild, "Earth

Science: Life Battered but Unbowed," *Nature* 459 (May 21, 2009), pp. 335–36 (doi:10.1038/459335a); and Simon A. Wilde et al., "Evidence from Detrital Zircons for the Existence of Continental Crust and Oceans on the Earth 4.4 Gyr Ago," *Nature* 409 (January 11, 2001), pp. 175–78 (doi:10.1038/35051550).

One of the key studies that showed that the moon was blasted off the very young Earth by an impact from another large celestial body, Theia, the Titan of ancient Greek myth who gave birth to the Moon goddess, Selene, is: U. Wiechert et al., "Oxygen Isotopes and the Moon-Forming Giant Impact," *Science* 294 (October 12, 2001), pp. 345–48 (doi:10.1126/science.1063037).

On Darwin's handwriting: as part of the filming of BBC4's *The Cell*, I had the great privilege to hold in my ungloved hands the letter Darwin sent to Joseph Hooker in 1871 in which he ponders his now-famous "warm little pond." Standing in a reading room in the University of Cambridge Library, I read an extract from this priceless piece of history. After several takes were spoiled because I couldn't decipher his scrawl, we ended up slipping a printed transcript on top of the actual letter. It is, of course, entirely irrelevant that his handwriting was so awful, but I find this quite amusing.

J. B. S. "Jack" Haldane is one of the dominant figures of twentieth-century biology in many of its forms, and a truly fascinating chap. Not only did he ponder the origin and nature of life and species, he was a central figure in the fusion of genetics and evolutionary biology and many other fundamental aspects of the living world, and back in 1923 he was probably the first person to suggest hydrogen as a basis for renewable energy. He was a Marxist and vocally spoke out against the British government's role in the Suez Crisis in 1956. His classic essay "On Being the Right Size" is available for free online, as is a very curious film presented by Haldane (produced by the Soviet Film Agency, 1940) entitled *Experiments in the Revival of Organisms*, which features the posthumous physical reactions of a severed dog's head. Unfortunately, the definitive biography by Ronald Clark, *The Life and Works of J. B. S. Haldane* (1968), is currently out of print.

This is Stanley Miller's iconic experiment from 1953: S. L. Miller, "A Production of Amino Acids under Possible Primitive Earth Conditions," *Science* 117 (May 15, 1953), pp. 528–29 (doi:10.1126/science.117.3046.528).

And Jeffrey Bada's 2008 analysis of some of Miller's experiments: Adam P. Johnson et al., "The Miller Volcanic Spark Discharge Experiment," *Science* 322 (17 October 2008), p. 404 (doi:10.1126/science.1161527).

Chapter 4: What Is Life?

A good read on the accumulation of information storage and inheritance in molecules is: G. F. Joyce, "Bit by Bit: The Darwinian Basis of Life," *PLoS Biology* 10 (2012) (e1001323. doi:10.1371/).

About various definitions of life: Pier Luigi Luisi, *Origins of Life and Evolution of the Biosphere* 28 (1998), pp. 613–22.

Refer to Edward Trifinov's meta-definition of life, based on the words that scientists use. Nineteen retorts can be found in the February 2012 edition of the same journal: Edward N. Trifonov, "Vocabulary of Definitions of Life Suggests a Definition," *Journal of Biomolecular Structure and Dynamics* 29 (October 2011), pp. 259–66.

Both Erwin Schrödinger's and J. B .S. Haldane's books entitled *What Is Life?* (1944 and 1949, respectively) are available online for free, and are both essential reading.

Chapter 5: The Origin of the Code

The theoretical takeover of DNA from RNA in the first genes: Kevin Leu et al., "On the Prebiotic Evolutionary Advantage of Transferring Genetic Information from RNA to DNA," *Nucleic Acids Research* 39 (2011), pp. 8135–47 (doi:10.1093/nar/gkr525).

On Jerry Joyce and Tracey Lincoln's short RNA molecules that have the effect of endlessly reproducing themselves: Tracey A. Lincoln and Gerald F. Joyce, "Self-Sustained Replication of an RNA Enzyme," *Science* 32 (February 27, 2009), pp. 1229–32 (doi:10.1126/science.1167856).

On Jack Szostak and David Bartel's classic ribozyme experiment: D. P. Bartel and J. W. Szostak, "Isolation of New Ribozymes from a Large Pool of Random Sequences," *Science* 261 (September 10, 1993), pp. 1411–18.

Admittedly a long way from infinite monkeys, but half a dozen aren't very Shakespearean: David Adam, "Give Six Monkeys a Computer, and What Do You Get? Certainly Not the Bard," *Guardian* (May 9, 2003).

On the idea of a ribozyme's being the most distant ancestor inspired by the blog *Tales from the Nobel Factory,* written by Alex Taylor: though I have chosen a different ribozyme as a conceptual aboriginal ancestor, I am most grateful to him for this and other bits he helped me with, though not his inability to understand what a collective noun is: http://talesfromthenobelfactory.posterous .com; Aniela Wochner et al., "Ribozyme-Catalyzed Transcription of an Active Ribozyme," *Science* 332 (April 8, 2011), pp. 209–12 (doi:10.1126/science .1200752).

Three papers on the origin of the genetic code with fewer than the four bases of modern life bear reading: Jeff Rogers and Gerald F. Joyce, "A Ribozyme That Lacks Cytidine," *Nature* 402 (November 18, 1999), pp. 323–25 (doi:10 .1038/46335); John S. Reader and Gerald F. Joyce, "A Ribozyme Composed of Only Two Different Nucleotides," *Nature* 420 (December 19, 2002), pp. 841–44 (doi:10.1038/nature01185); Julia Derr et al., "Prebiotically Plausible Mechanisms Increase Compositional Diversity of Nucleic Acid Sequences," *Nucleic Acids Research* (2012) (doi:10.1093/nar/gks065).

John Sutherland's classic synthesis of uracil: Matthew W. Powner, Béatrice Gerland, and John D. Sutherland, "Synthesis of Activated Pyrimidine Ribonucleotides in Prebiotically Plausible Conditions," *Nature* 459 (May 14, 2009), pp. 239–42 (doi:10.1038/nature08013).

This fantastic study demonstrates that the uracil found in the Murchison Meteorite was from space, and not earthly contamination: Zita Martins et al., "Extraterrestrial Nucleobases in the Murchison Meteorite," *Earth and Planetary Science Letters* 270 (2008), pp. 130–36.

Chapter 6: Genesis

On the accumulation of human culture: Adam Powell, Stephen Shennan, and Mark G. Thomas, "Late Pleistocene Demography and the Appearance of Modern Human Behavior," *Science* 324 (June 5, 2009), pp. 1298–1301 (doi:10.1126/science.1170165).

On competition in cell membrane formation: Itay Budin and Jack W. Szostak, "Physical Effects Underlying the Transition from Primitive to Modern Cell Membranes," *PNAS* (2011) (doi:10.1073/pnas.1100498108).

Jack Szostak's protocell model dominates studies of the emergence of the membrane. Other ideas and experiments have suggested different routes. One recent study by Stephen Mann, Shogo Koga, and colleagues from the University of Bristol puts forth the possibility of microdroplets stuffed with nucleotides and amino acid strings so that they compartmentalize without a physical skin. We await the development of that idea with relish. Shogo Koga et al., "Peptide-nucleotide Microdroplets as a Step towards a Membrane-free Protocell Model," *Nature Chemistry* 3 (2011), pp. 720–24 (doi:10.1038/nchem.1110).

Here's an interesting study from 2011 that shows replication of DNA within self-reproducing protocells and sharing of the DNA between the daughter cells. Although very clearly not the same as in modern cells, the process mimics the outcome of actual cell division: K. Kurihara et al., "Self-reproduction of Supramolecular Giant Vesicles Combined with the Amplification of Encapsulated DNA," *Nature Chemistry* 3 (2011), pp. 775–81 (doi:10.1038/nchem.1127).

Orgel's thoughts on the RNA World: Leslie E. Orgel, "Prebiotic Chemistry and the Origin of the RNA World," *Critical Reviews in Biochemistry and Molecular Biology* 39 (2004), pp. 99–123 (doi:10.1080/10409230490460765).

A 2001 overview of the emerging experiments to emulate abiogenesis: Jack W. Szostak, David P. Bartel, and P. Luigi Luisi, "Synthesizing Life," *Nature* 409 (January 18, 2001), pp. 387–90 (doi:10.1038/35053176).

The first description of the hydrothermal vents of the Lost City: D. S. Kelly et al., "An Off-axis Hydrothermal-vent Field near the Mid-Atlantic Ridge at 30° N," *Nature* 412 (July 12, 2001), pp. 145–49.

If you can get hold of this paper, it is a hoot: Crick and Orgel pondering aliens as the origin of earthly life: F. H. C. Crick and L. E. Orgel, "Directed Panspermia," *Icarus* 19 (1973), pp. 341–46.

Sir Fred Hoyle and Chandra Wickramasinghe's interesting but wrong ideas on panspermia: Sir Fred Hoyle and Chandra Wickramasinghe, *Evolution from Space: A Theory of Cosmic Creationism* (Touchstone, 1984).

There are a handful of papers and reviews by Nick Lane, Mike Russell, Bill Martin and others on proton gradients, and the energetic origin of life in hydrothermal vents. Here are some of the best: N. Lane, W. Martin, "The origin of membrane bioenergetics," *Cell* 151 (2012), pp. 1406–16; W. Martin, "Hydrogen, metals, bifurcating electrons, and proton gradients: the early evolution of biological energy conservation," *FEBS Letters* 586 (2012), pp. 485–93; M. J. Russell (editor), *Origins, Abiogenesis and the Search for Life in the Universe* (Cosmology Science Publishers: Cambridge, MA, 2011), p. 487; N. Lane, J. F. Allen, and W. Martin, "How did LUCA make a living? Chemiosmosis in the origin of life," *Bioessays* 32 (2010), pp. 271-280; W. Martin, J. Baross, D. Kelley, and

262   Annotated Bibliography

M. J. Russell, "Hydrothermal vents and the origin of life," *Nature Reviews: Microbiology* 6 (2008), pp. 806–814; W. Martin, M. J. Russell, "On the origin of biochemistry at an alkaline hydrothermal vent," *Philosophical Transactions*, Royal Society of London (Ser.B) 362 (2007), pp. 1887–1925.

And this is a profile of Mike Russell and his research: John Whitfield, "Origin of life: Nascence man," *Nature* 459 (2009), pp. 316–319 (doi:10.1038/459316a).

Part II: The Future of Life

Chapter 7: Life, Not as We Know It

Craig Venter's paper on building a cell from a fully manufactured genome: Daniel G. Gibson et al., "Creation of a Bacterial Cell Controlled by a Chemically Synthesized Genome," *Science* 329 (July 2, 2010), pp. 52–56 (doi:10.1126/science.1190719).

And the patent for the minimal genome of *Mycoplasma mycoides*, or Synthia, as it was nicknamed: http://1.usa.gov/QsssVZ.

Chapter 8: Created, Not Begotten

I first physically encountered Freckles the spider-goat (and her creator, Randy Lewis) during the filming of an episode of the BBC television science series *Horizon*. Parts of this chapter are taken from interviews from that filming, and from an article I wrote for the *Observer* newspaper based on that show: "Synthetic Biology and the Rise of the 'Spider-goats'" (January 14, 2012).

Here is one of many papers describing the silk-producing genetic modifications: F. Teulé et al., "A Protocol for the Production of Recombinant Spider Silk-like Proteins for Artificial Fiber Spinning," *Nature Protocols* 4:3 (2009), pp. 341–55 (doi:10.1038/nprot.2008.250).

Here is Hamilton Smith's 1970 paper characterizing the restriction enzyme HinDII; it was this technology that enabled all molecular biology that followed: H. Smith and K. W. Wilcox, "A Restriction Enzyme from Hemophilus Influenzae 1. Purification and General Properties," *Journal of Molecular Biology* 51 (1970), pp. 379–91 (doi:10.1016/0022-2836(70)90149-X).

Chapter 9: Logic in Life

The most comprehensive analysis of modern biotechnology is *Biology Is Technology: The Promise, Peril, and New Business of Engineering Life* by Robert H. Carlson (Harvard University Press, 2011), and is a brilliant primer for the science, economics, and politics of this ever-changing field.

Ron Weiss's cancer assassin circuit: Zhen Xie et al., "Multi-Input RNAi-Based Logic Circuit for Identification of Specific Cancer Cells," *Science* 333 (September 2, 2011), pp. 1307–11 (doi:10.1126/science.1205527).

The cancer cells that are targeted and destroyed in this system are called HeLa cells. Due to their unlimited ability to divide—"immortality"—they have been experimented on and distributed innumerable times, and are of immeasurable importance to many aspects of modern biology. They were derived in 1951 from the cervix of Henrietta Lacks, a poor black American who died of her cancer and lies in an unmarked grave. Her tale is told with grace and brilliance by Rebecca Skloot in the book *The Immortal Life of Henrietta Lacks* (Crown, 2010).

The two back-to-back papers that launched the component nature of genetic engineering at the birth of synthetic biology: Michael B. Elowitz and Stanislas Leibler, "A Synthetic Oscillatory Network of Transcriptional Regulators," *Nature* 403 (January 20, 2000), pp. 335–38 (doi:10.1038/35002125); Timothy S. Gardner, Charles R. Cantor, and James J. Collins, "The Construction of a Genetic Toggle Switch in *Escherichia coli*," Nature 403 (January 20, 2000), pp. 339–42 (doi:10.1038/35002131).

Ten years later, genetic circuits had become significantly more sophisticated. For example: Tal Danino, Octavio Mondragón-Palomino, Lev Tsimring, and Jeff Hasty, "A Synchronized Quorum of Genetic Clocks," *Nature* 463 (January 21, 2010), pp. 326–30 (doi:10.1038/nature08753).

At *Nature,* my colleague Charlotte Stoddart made a short film of the astonishing synthetic synchronization of Hasty's bacteria and published it on YouTube. In it Charlotte's voice-over describes the fluorescing action as a "Mexican wave," because it closely resembles the metachronal rhythm demonstrated in packed stadiums by sports fans standing, yelling, and sitting down in succession. YouTube viewers are not known for providing the most sophisticated commentary below the videos themselves, but we were surprised to learn that the term *Mexican wave* was primarily a British phrase, as a result of its occurrence at football (soccer) matches in the Mexico World Cup, 1986. It seems that many other people in the world took this as a vaguely racist comment.

On the dangers of a trip to Mars: W. Friedberg et al., "Health Aspects of Radiation Exposure on a Simulated Mission to Mars," *Radioactivity in the Environment* 7 (2005), pp. 894–901.

A good analysis of the claims and hype of synthetic biology: Roberta Kwok, "Five Hard Truths for Synthetic Biology," *Nature* 463 (January 20, 2010), pp. 288–90 (doi:10.1038/463288a).

Chapter 10: Remix and Revolution

My thoughts on copying, creativity, and music in relation to evolution and synthetic biology are influenced, appropriately enough, by a series of online videos by New York filmmaker Kirby Ferguson entitled *Everything Is a Remix,* available in four parts at http://www.everythingisaremix.info/.

Completing the loop, Ferguson used evolution as an introductory metaphor for the genesis of ideas in part 4 of his outstanding piece of work, which he formulated with a very small amount of help from me. I now gleefully extend this idea into the realm of synthetic biology by borrowing and transforming his ideas. I'm pretty sure he won't mind.

The changing nature of creativity, specifically in music, as a result of copyright law is excellently documented by Joanna Demers in *Steal This Music* (University of Georgia Press, 2006).

One of the most striking examples of sampling in music is the innumerable reuses of a short stretch of drum solo from the B-side of a 1969 7-inch single by the Winstons, a largely forgotten funk-soul band. The four-bar section is known as the Amen break, as the song itself was called "Amen, Brother," and features a dropped and syncopated drum hit known as a breakbeat. In the mid-1980s, it began appearing as samples in tracks on the developing New York hip-hop scene. By the end of the 1980s, it had been sampled widely, by many dance and hip-hop bands, including in "Straight Outta Compton" by the influential group N.W.A. A decade later, the seven-second drum loop had appeared on hundreds of songs ranging from a David Bowie single to the credits of the cartoon *Futurama*. Entire genres of jungle music and drum and bass are based around the Amen break. A database of tracks that use it can be found at http://amenbreakdb.com/.

The first album to be constructed entirely from samples is *Endtroducing* (1996) by DJ Shadow, which uses at least ninety-nine samples over the eighteen tracks, from films including *Silent Running* (1972) and *Prince of Darkness* (1987), and bands including Queen, Metallica, Pink Floyd, Kraftwerk, and Nirvana. Craig Venter's bacteria nicknamed Synthia is probably the closest to a fully sampled living cell, though *Endtroducing* is more culturally relevant and much more exciting.

The edge-detecting synthetic bacteria that spelled out the simple coder's test message "Hello world": Anselm Levskaya et al., "Synthetic Biology: Engineering *Escherichia coli* to See Light," *Nature* 438 (November 24, 2005), pp. 441–42 (doi:10.1038/nature04405).

Three months of measuring the wings and eyes of several thousand stalk-eyed flies (*Cyrtodiopsis dalmanni*), after breeding and raising them in liquefied rotting sweet corn—happy days: P. David, A. Hingle, D. Greig, A. Rutherford, A. Pomiankowski, and K. Fowler, "Male Sexual Ornament Size but Not Asymmetry Reflects Condition in Stalk-eyed Flies," *Proceedings of the Royal Society B* 265 (1998), p. 2211 (doi:10.1098/rspb.1998.0561).

The beginnings of the biocementation process that the Brown-Stanford iGem team would use to look at RegoBrick formation for terraforming: S. S. Bang and V. Ramakrishnan, "Calcite Precipitation Induced by Polyurethane-immobilized *Bacillus pasteurii*," *Enzyme and Microbial Technology* 28 (2001), pp. 404–9.

Two useful reviews of patents in genetics and synthetic biology: Arti Rai and James Boyle, "Synthetic Biology: Caught between Property Rights, the Public Domain, and the Commons," *PLoS Biology* 5 (2007), p. e58 (doi:10.1371/journal.pbio.0050058); Berthold Rutz, "Synthetic Biology and Patents: A European Perspective," *EMBO Reports* 10 (2009), pp. S14–S17 (doi:10.1038/embor.2009.131).

The European patent for the OncoMouse: http://register.epoline.org/espacenet/application?number=EP85304490.

President Obama's *National Bioeconomy Blueprint*: http://www.whitehouse
.gov/sites/default/files/microsites/ostp/national_bioeconomy_blueprint_april_
2012.pdf.

Chapter 11: The Case for Progress

The science writer Martin Robbins, along with many supporters of the scientists, attended Take the Flour Back's planned attack on the Rothamsted GM
fields, "'HULK SMASH GM'—Mixing Angry Greens with Bad Science," May
30, 2012, www.guardian.co.uk.

Here are several papers on specific aspects of genetic modification that are
cited in the first few pages of chapter 11: Sophie Vandermoten, "Aphid Alarm
Pheromone: An Overview of Current Knowledge on Biosynthesis and Func-
tions," *Insect Biochemistry and Molecular Biology* 42 (2012), pp. 155–63;
G. Kunert, C. Reinhold, and J. Gershenzon, "Constitutive Emission of the
Aphid Alarm Pheromone, (E)-β-farnesene, from Plants Does Not Serve as a
Direct Defense Against Aphids," *BMC Ecology*, 10 (2010), p. 23 (doi:10.1186/
1472-6785-10-23); L. G. Firbank et al., "Farm-Scale Evaluation of Genetically
Modified Crops," *Nature* 399 (1999), pp. 727–28; J. N. Perry et al., "Ban on Tri-
azine Herbicides Likely to Reduce but Not Negate Relative Benefits of GMHT
Maize Cropping," *Nature* 428 (18 March 2004), pp. 313–16 (doi:10.1038/na-
ture02374); M. S. Heard et al., "Weeds in Fields with Contrasting Conventional
and Genetically Modified Herbicide-Tolerant Crops. 1. Effects on Abundance
and Diversity," *Philosophical Transactions of the Royal Society B* 358 (2003), pp.
1819–32; J. N. Perry et al., "Design, Analysis and Power of the Farm-Scale Eval-
uations of Genetically Modified Herbicide-Tolerant Crops," *Journal of Applied
Ecology* 40 (2003), pp. 17–31; Chelsea Snella et al., "Assessment of the Health
Impact of GM Plant Diets in Long-term and Multigenerational Animal Feeding Trials: A Literature Review," *Food and Chemical Toxicology* 50 (2012), pp.
1134–48;

Take the Flour Back's website: http://taketheflourback.org/.

The Presidential Commission for the Study of Bioethical Issues' recommendations on synthetic biology, 2010: "New Directions: The Ethics of Synthetic Biology and Emerging Technologies," http://bioethics.gov/cms/
synthetic-biology-report.

Also, the response from Friends of the Earth and the ETC: http://www.foe
.org/news/blog/2010-12-groups-criticize-presidential-commissions-
recommenda.

The Obama administration's *National Bioeconomy Blueprint*: http://www
.whitehouse.gov/blog/2012/04/26/national-bioeconomy-blueprint-released.

And the response from Friends of the Earth and the ETC: "Principles for
the Oversight of Synthetic Biology," http://www.foe.org/projects/food-and
-technology/blog/2012-03-global-coalition-calls-oversight-synthetic-biology.

This classic paper was the first real instance of genetic modification, by Paul
Berg and his team: D. A. Jackson, R. H. Symons, and P. Berg, "Biochemical
Method for Inserting New Genetic Information into DNA of Simian Virus 40:

Circular SV40 DNA Containing Lambda Phage Genes and the Galactose Op-
eron of *Escherichia coli*," *Proceedings of the National Academy of Science USA*
69 (1972), pp. 2904–9.

The post-Asilomar schism reported: "Environmental Groups Lose Friends
in Effort to Control DNA Research," *Science* 202 (1978), p. 22.

ETC's analysis of the bio-based economy: "The New Biomasters: Synthetic
Biology and the Next Assault on Biodiversity and Livelihoods," http://www
.etcgroup.org/content/new-biomassters.

The Guardian's stunt to assemble parts of the smallpox genome by mail
order: James Randerson, "Revealed: The Lax Laws That Could Allow Assembly
of Deadly Virus DNA," *Guardian*, June 14, 2006.

The 2002 construction of the virus that causes polio by taking its genome
sequence from publicly available databases: Jeronimo Cello, Aniko V. Paul, and
Eckard Wimmer, "Chemical Synthesis of Poliovirus cDNA: Generation of Infectious Virus in the Absence of Natural Template," *Science* 297 (August 9,
2002), pp. 1016–18 (doi:10.1126/science.1072266).

And of the 1918 flu: Terrence M. Tumpey et al., "Characterization of the
Reconstructed 1918 Spanish Influenza Pandemic Virus," *Science* 310 (2005),
pp. 77–80 (doi:10.1126/science.1119392).

The two experimental flu papers, one by Kawaoka in *Nature* and the other
by Fouchier in *Science*, both finally published in full in June 2012 after much
deliberation and debate, and an earlier editorial from *Nature*: Masaki Imai et
al., "Experimental Adaptation of an Influenza H5 HA Confers Respiratory
Droplet Transmission to a Reassortant H5 HA/H1N1 Virus in Ferrets," *Nature*
486 (June 21, 2012), pp. 420–28 (doi:10.1038/nature10831); Sander Herfst et al.,
"Airborne Transmission of Influenza A/H5N1 Virus Between Ferrets," *Science*
336 (June 22, 2012), pp. 1534–41 (doi:10.1126/science.1213362); "Publishing
Risky Research," *Nature* 485 (May 3, 2012), p. 5 (doi:10.1038/485005a).

Though somewhat dated, this is a useful review of actions for biologists to
take to restrict bioterrosim: Claire M. Fraser and Malcolm R. Dando, "Genomics and Future Biological Weapons: The Need for Preventive Action by the Bio-
medical Community," *Nature Genetics* 29 (2001), pp. 253–56 (doi:10.1038/
ng763).

Gilles-Eric Séralini's controversial study on the negative effects of GM crops
in rats' diets: Gilles-Eric Séralini et al., "Long term Toxicity of a Roundup Herbicide and a Roundup-Tolerant Genetically Modified Maize," *Food and Chemical Toxicology* 50 (2012), pp. 4221–31.

And two months later, a damning review by the European Food Safety Authority: European Food Safety Authority, "Final Review of the Séralini et al.
(2012a) Publication on a 2-year Rodent Feeding Study with Glyphosate Formu-
lations and GM Maize NK603 as Published Online on 19 September 2012 in
Food and Chemical Toxicology," *EFSA Journal* 10 (2012), p. 2986.

On the creation of artemisinin: Dae-Kyun Ro et al., "Production of the Antimalarial Drug Precursor Artemisinic Acid in Engineered Yeast," *Nature* 440
(April 13, 2006), pp. 940–43 (doi:10.1038/nature04640); P. J. Westfall et al.,
"Production of Amorphadiene in Yeast, and Its Conversion to Dihydroarte-
misinic Acid, Precursor to the Antimalarial Agent Artemisinin," *Proceedings*

of the National Academy of Science USA 109 E111-8 (January 12, 2012); Declan Butler, "Malaria Drug-makers Ignore WHO Ban," *Nature* 460 (July 14, 2009), pp. 310–11 (doi:10.1038/460310b); Melissa Lee Phillips, "Genome Analysis Homes In on Malaria-Drug Resistance," *Nature News* (April 5, 2012) (doi:10.1038/nature.2012.10398).

The United Kingdom's chief scientific adviser on GM foods: "GM Food Needed to Avert Global Crisis, Says Government Adviser," *Telegraph* (January 24, 2011).

"Meanings of 'Life,'" *Nature* 447 (June 28, 2007), pp. 1031–32 (doi:10.1038/4471031b).

Written by some of the giants of the first wave of genetic modification, the recommendations following the Asilomar summit. Paul Berg, David Baltimore, Sydney Brenner, Richard O. Roblin III, and Maxine F. Singer, "Summary Statement of the Asilomar Conference on Recombinant DNA Molecules," *Proceedings of the National Academy of Science USA* 72 (1975), pp. 1981–84.

A major public assessment of how various publics view synthetic biology in the United Kingdom: "BBSRC Synthetic Biology Dialogue," www.bbsrc.ac.uk/web/FILES/Reviews/synbio_summary-report.pdf.

Afterword

Babel, Esperanto, Klingon, Babm, Blissymbolics, Loglan, and Lojba—see Arika Okrent's wonderful book on the stories of how we have started nearly nine hundred new languages: *In the Land of Invented Languages: Esperanto Rock Stars, Klingon Poets, Loglan Lovers, and the Mad Dreamers Who Tried to Build a Perfect Language* (Spiegel & Grau, 2009).

On Steve Benner's addition of Z and P to the natural bases A, T, C, and G: Z. Yang, F. Chen, et al., "Amplification, Mutation, and Sequencing of a Six-Letter Synthetic Genetic System," *Journal of the American Chemical Society* (2011) (doi:10.1021/ja204910n).

On XNA and the birth of synthetic genetics: Vitor B. Pinheiro et al., "Synthetic Genetic Polymers Capable of Heredity and Evolution," *Science* 336 (April 20, 2012), pp. 341–44 (doi:10.1126/science.1217622).

On the creation of unnatural amino acids: Lloyd Davis and Jason W. Chin, "Designer Proteins: Applications of Genetic Code Expansion in Cell Biology," *Nature Reviews Molecular Cell Biology* 13 (February 2012), pp. 168–82 (doi:10.1038/nrm3286); Jason W. Chin et al., "Addition of a Photocrosslinking Amino Acid to the Genetic Code of *Escherichia coli*," *PNAS* 99 (2002), pp. 11020–24 (doi:10.1073/pnas.172226299).

On ancient DNA: one of the most surprising interviews I have done, for the *Nature Podcast,* was with Stephan C. Schuster, the lead researcher on the sequencing of the mammoth genome, on November 20, 2008:

SCHUSTER: Once we realized that there might be the possibility that the hair shaft contained actual DNA, we went out and tried to find sources of DNA that we could reliably step into. And the way I do this, I started searching for them on eBay and when I immediately found that there are zillions of hairs

available, we contacted the seller and then, together with au-
thorities of the university, we made sure that there were
proper import permits, and we also had paleontologists and
museum curators in Russia to check on the sources of that
because we were very worried that some of those hairs and
fossils might have illegally been sold outside Russia. And after
we verified that all of this was taken care of and there is a
clean record, then we started buying hair in a larger supply
from that source.

AR: Some of these details aren't mentioned in the methods section
of the paper. Can I just get you to say that again: You bought
these hairs off eBay? What was the successful bid?

SCHUSTER: For a handful of hair, I think it cost me something like
132 bucks.

On the wooly mammoth genome sequence: Webb Miller et al., "Sequencing
the Nuclear Genome of the Extinct Wooly Mammoth," *Nature* 456 (November
20, 2008), pp. 387–90 (doi:10.1038/nature07446).

An assessment of sequencing DNA from long-dead specimens: S. Pääbo et
al., "Genetic Analyses from Ancient DNA," *Annual Review of Genetics* 38
(2004), pp. 645–79 (doi:10.1146/annurev.genet.37.110801.143214).

On very old human DNA: Richard E. Green et al., "A Draft Sequence of the
Neanderthal Genome," *Science* 328 (May 7, 2010), pp. 710–22 (doi:10.1126/sci-
ence.1188021).

On exceedingly old plant DNA: Eske Willerslev et al., "Ancient Biomolecules from Deep Ice Cores Reveal a Forested Southern Greenland," *Science* 317
(July 6, 2007), pp. 111–14 (doi: 10.1126/science.1141758).

One of the first suggestions in the academic literature that DNA could be
used as a digital storage device: Eric B. Baum, "Building an Associative Memory Vastly Larger Than the Brain," *Science* 268 (April 28, 1995), pp. 583–85
(doi:10.1126/science.7725109).

The most advanced realization of DNA as a means of storing data: George
M. Church, Yuan Gao, and Sriram Kosuri, "Next-Generation Digital Information Storage in DNA," *Science* 337 (September 28, 2012) (doi:10.1126/science
.1226355).

Index

274   Index